# HYDROPONICS FOR BEGINNERS

**The Complete Step by Step Guide on how to build your Hydroponic System and Start Growing herbs and vegetables without Soil, at home and in your Greenhouse**

**Michael Nestor**

# Table of Contents

# INTRODUCTION

More and more people these days are realizing that the above statement is true. Hydroponics is the best, cleaner and more controlled way to grow. From your prize flowers' blooms to the sparkling fresh taste of the salad you just picked to the satisfying taste of your fresh vegetables picked and eaten the same day. I say, you can do it better with hydroponics.

## WHAT IS HYDROPONICS?

There is nothing magic or mysterious about hydroponic garden growing. Hydroponic gardening is merely a method of removing or replacing the usual soil with a clean substitute such as rock wool, coco grow, peat, clay pebbles, etc.

In the method of soil gardening the plant's roots dig and search into the soil for their needed nutrients whereas in hydroponic growing the exact nutrients designed to each plant's requirements are supplied and controlled in an automatically timed feeding.

It follows that as the plant doesn't have to work as hard for their nutrients the results are more energy being supplied to produce lush blooms for flowers and healthier more abundant vegetable crops.

## WHY NOT SOIL GARDENING?

So often these days' soil has become depleted from over use and contaminated with harmful chemicals. In other words, not dependable for healthy produce. With hydroponic growing there is no worries about lack of nutrition in your hydroponic grown vegetables and fruit and no worries about you or your family consuming pesticide or other chemical residues while you are eating the food. The only thing that gets into your hydroponic grown food is what you put into it.

A small easy to use complete hydroponic system for home use is easy to get and quite affordable and that is why this book was written, too. Easy and fun to set up, the systems come complete with all that you need to get started together with all the how to instructions and guidelines for success. Quality manufacturers all over the world are standing by ready to help in any way they can with technical support or parts replacements if needed.

An indoor hydroponic garden can easily become a child's or family project. How about a science project?

## WHAT KIND SHOULD I GET?

That depends solely on how much and what you want to grow? There are a wide delightful variety of hydroponic indoor garden systems to choose from. The systems come in different sizes from the small 6 planter such as "Emily's Garden System" all the way to the

"Aeroflo2 - 60" which is expandable from 60 to 120 plant sites. Some of the "Aeroflo" series are long and narrow and fit easily along a window for the natural sunlight. Then there is the "Ebb Monster" which is used for growing berries or fruit trees or larger vegetable varieties. There are some systems which double for plant cloning and/or plant cultivation.

So, you see, whatever you want to grow: whether it be carrots, tomatoes, kale, flowers, citrus, berries, peppers, the hydroponics people have you covered.

Whatever your desires or goals there are systems to go along with it.

**NOW YOU HAVE IT! WHERE WILL YOU PUT IT?**

Some people think closet growing works best for them. It's relatively easy to control lighting and temperature in a small closet area. No drafts, etc.

Some, like me, prefer their garden out in the open to watch and enjoy maybe as a focal point as you would place or use a pretty house plant. On the top of an end table or corner table or dining room table. Some people have a complete vegetable garden set up inside their garage. There are small indoor greenhouses available and affordable that work well in a garage environment: controlling temperature, humidity and light.

**WHAT ABOUT LIGHT?**

Outdoor growing, of course, depends entirely upon the sunlight. Indoors it works well also if you have placed your hydroponic system near a window or in a well-lit inside porch. If not you could consider something like the affordable and efficient Jump Start Lighting System which comes in two convenient sizes, 2 feet and 4 feet in length, and is entirely portable with a T5 fluorescent light bulb that can simulate the sun.

## NUTRIENTS! HOW DO I KNOW WHICH TO USE?

Easy! Nutrients are usually included with each new system along with amounts guidance. The manufacturers of the nutrients offer much instructional guidance and how-to. There are wonderful liquid plant nutrients available all the way from encouraging your flowers to have larger more frequent and colorful blooms to helping your fruits and vegetables blossom earlier and more abundantly. There are nutrients to make your berries, grapes and citrus taste sweeter.

## WHERE WILL YOU GET YOUR PLANTS TO PLACE IN YOUR NEW INDOOR HYDROPONIC GARDEN?

You can always grow them from seeds as usual. There is help out there for that also. A Germination Station, or Hot House Plus is a great help in the sprouting of

seeds. Most of the seedling/germination kits have heat mats and humidity domes to make sure of your success by making the ideal environment for your seedlings to increase their growth rate.

Or if you want to skip the seedling part of the deal you can get a new plant directly from rooting cuttings of your neighbor's or friend's thriving plants.

There is help out there also for that. The propagation or cloning machines, or aeroponic systems are available, affordable and easy to get online with instructional guidelines and how-to to get you started in the right direction. They are easy to use. You just place your cuttings in the small holes provided in your system and beneath the covered area there is timed automatic spraying of the cuttings with either a plant nutrient water mix or oxygenated water such as the method used in the "OxyClone" System.

Either way they are no bother and an easy success.

So, you see, Its all out there and ready for you to have.

Think of the adventure of it all! The fun!

"Yes, thank you, they are delicious, aren't they? I just picked them today from my garden," you'll be saying soon at your family dinner or dinner party. A small smug smile is allowed at this time.

## TYPES OF THE MOST USED HYDROPONIC SYSTEMS

We've all been taught that for any plant to grow it needs soil. But the science of hydroponics discards that theory. Hydroponics is the way of growing plants without the need of soil. The main purpose of soil is to hold the nutrients for the plants. If the nutrients can be accessed directly, then the soil can be eliminated as it serves no purpose. In cases where mediums are needed to hold the nutrient solutions temporarily, mineral wool or gravel can be used.

Minerals can be diluted in water and the roots of the plants can be immersed in it. This allows for more yields per acreage. The consumption of water is reduced by almost 60%, as there is no waste. Also, since the nutrients are artificially introduced, hydroponics does not cause the depletion of nutrients in the soil.

### Types of Hydroponics

There are six types of hydroponics methods. These are:

1. Deep Water Culture (DWC): In this method, the plants roots are suspended directly into the solution of

nutrients in oxygenated water. The roots are super oxygenated with the combination of an air pump and porous stones. The growth rate of these plants is improved because of the high amount of oxygen the roots receive.

2. Flood and Drain (Ebb and Flow): Here the plants are planted in a medium that is in a tray. The tray in turn is suspended over a reservoir of the mineral and nutrients solution. A pump, at intervals, pumps the solution into the tray and floods the medium. When the desired level of flooding is achieved the pump switches off. The process starts again after the solution has drained back down into the reservoir.

3. Nutrient Film Technique: This is a technique where the roots of the plants, which are suspended in a watertight gulley, are continuously washed in a circulating stream of water that contains the minerals and nutrients. Since the roots of the plants form a mat on the bottom of the gulley, the circulating water is just enough to cover the roots; hence the term 'Water Film.'

4. Wick System: In this technique all the plants are in a container and in a growing medium. An absorbent wick, usually a nylon rope, is buried partially in each container. One end of the rope is left dangling in the mineral and nutrient solution. The wick pulls the solution up and into the growing medium thus feeding the roots.

5. Drip System: Here again, a pump is required. All the plants are in their own trays and separated from the nutrients and minerals solution. The pump then pumps the solution and feeds the individual plant, drip by drip, from above. The drip rate can be controlled to either go faster or slower depending on the plants needs.

6. Aeroponics: In this sixth, and final, method the plants are suspended in a box or container and their roots are exposed. A pump, at constant intervals, pumps a fine mist of the nutrient solution at the roots. The main advantage of this method is excellent aeration for the roots.

# HYDROPONICS VERSUS SOIL

To understand what the advantages are growing with hydroponics, first you must understand what hydroponics is. Hydroponics is simple. Hydroponics is growing with water instead of soil. Typically a user adds concentrated nutrients into the water, simulating the fertilizers found in soil.

With hydroponics you have much more control over your grow then you do with soil, because simply adding the right amount of nutrients in the water guarantees that you will have the right amount of food for your plants. In soil it can be much more difficult to diagnose when you are short mineral trace elements, because you really don't know what was missing to begin with.

However when using hydroponics nutrients are pre-mixed with exactly the right amount of trace elements. So by simply adding the proper amount of hydroponic nutrients to your water in the hydroponic system, you know that the nutrients have been mixed correctly and the plants have all the elements they need to survive. If your plants should never begin to get sick such as yellowing leaves due to some sort of nutrient deficiency all you have to do is dump your hydroponic systems water, and fill it with fresh water and fresh nutrients.

Another huge advantage to hydroponics over soil is that in hydroponics you can grow your plants up to 10 times faster than soil. In soil your plants have to develop a very long and very integrated root system to get all the water and nutrients it needs out of the earth. Your plants need to develop a massive root system in the soil in order to get everything that's essential for them to grow quickly. So a plant grown in soil you would see a very large root system under the ground, but a very small plant above ground.

Hydroponics is exactly the opposite. Because it is so it easy for the plants to get the water and nutrients directly out of the water in your reservoir the plant can grow a much smaller root system and obtain the same amount of nutrients, if not more. In a hydroponic system you can expect to see a plant with a very small root system to get a very large plant above the ground. Essentially plants grow in hydroponics 10 times faster versus soil grow. This is a huge advantage. Plants grown in hydroponics versus soil is not only easier to grow, but they grow much faster as well.

Hydroponics is also very useful for people that live in parts of the country with extreme environment conditions. People living in a part of the country that is too hot, or too cold can make it very difficult to grow your plants. However, because hydroponics can be used indoors you can regulate the temperature by using either a heater or an air conditioner. Plants like to be at a comfortable temperature just like people do. If you can keep your plants right around 75 to 85° they will grow like wildfire all year. It would be impossible to try to maintain these conditions outdoors but it becomes very easy when using hydroponics. It is important that your plants do not get too hot or too cold during any part of the year or you will stunt their growth.

Hydroponics makes it quick and simple for anyone in the world to grow easily their own plants for food or for medicinal herbs. Hydroponics makes growing organics available to anybody with a source of fresh water and electricity. Hydroponics is even being developed by NASA by use of the space station. One day our astronauts will be eating food grown from a hydroponic system, just as you are about to embark on.

Hydroponics essentially gives your plant everything they want but spoon-fed, easy fashion making them grow very fast. If you were to enhance your hydroponic system by using $CO_2$, your plants can grow even double that speed. But that's another tutorial altogether.

## IS HYDROPONIC GARDENING BETTER FOR GROWING VEGETABLES?

So you want to know if hydroponic gardening is better for growing vegetables than conventional methods of gardening? Well, I suppose it is a matter of opinion. Some people have very busy lives and do not have the time to grow a garden, whether it is by conventional means or hydroponic gardening. Others do not have the space for a garden of any sort. For these people, purchasing store bought fresh vegetables is the way to go. However, the store bought vegetables and fruits are not fresh.

To people like me, fresh means straight from the garden. That is why I like hydroponic growing. I can grow all of my fruits and vegetables at home, even in my home, and be able to get them fresh whenever I want them. I can also have access to them at any time throughout the year. In the store, the produce is not grown hydroponically and certain items are only available at certain times of the year. If I can simulate the perfect growing conditions using hydroponic technique, I can get my favorite fruits and veggies at any time during the year. I do not have to wait for them to be "in season."

Hydroponic vegetables, veggies that are grown by using the hydroponic gardening technique, are generally better for a person's health. These hydroponic vegetables tend to be larger, juicier, and brighter in color than those found in the store. Store bought vegetables have a waxy film over them to keep them fresher for longer periods of time. Homegrown hydroponic vegetables have no film because there is no need for it. The waxy film is an additive put on by people to help with the transport and shelf life of the vegetables. Hydroponic vegetables are either eaten by you right away or still on the plant until you are ready to use them. So either way you look at it, hydroponic vegetables are better.

Vegetables are rich in antioxidants, which promote better health. Using hydroponic technique to grow your vegetables increases the antioxidant properties of your vegetables. This is because foods grown by using hydroponic technique are generally healthier than those grown by conventional methods of gardening. There is virtually little to no pests, so there is no need to poison the vegetables by spraying harmful and dangerous pesticides and insecticides. In a conventional garden, you have to worry about so many different bugs and pests that attack your plants. Most people resort to using pesticides from the store. Recent studies show that these pesticides, when ingested by humans, can be detrimental to a person's health and well-being.

So is hydroponic gardening better for growing vegetables? Like I said, it is a matter of opinion. I do know that this woman here will only accept home grown vegetables from hydroponic gardening. Only the best for me and mine. I fully believe that hydroponic gardening is the best for growing vegetables that not only look pretty but also taste so much better than the ones you buy from the store.

## AEROPONICS VS. HYDROPONICS VS. SOIL IS AGRICULTURAL PLANT SCIENCE

We all know what plants look like in soil and how soil provides mechanical stability. Plants absorb nutrients from the soil mix or as irrigation additives poured onto the soil. Plant roots work against the mechanical resistance of soil and expend energy searching for nutrients and oxygen in the soil. Water in moist soil helps the roots expand and with the absorption process at the root surface.

Hydroponics comes in varying levels where soil is replaced with nutrient in-solution water. Plant roots can be flooded and drained on a cycle, or in DWC hydroponics, permanently submerged in water (Deep Water Culture). Hydroponics offers the advantage of no energy wasted searching for nutrients.

Aeroponic systems are a specialized version of hydroponics where the roots of the plant extend only in air and the roots are directly sprayed with a nutrient water mix (the recipe). The primary difference is the availability of oxygen to the roots. In hydroponics, one has to be sure to supply oxygenated water. Standing water gets depleted of oxygen over time. In aeroponics, oxygen is surrounding the roots at all times. Surplus oxygen accelerates nutrient absorption at the root surface.

Plant support in both areoponics and hydroponics are provided by the hosting environment. Hydroponic plants tend to be stabilized with hydroton clay balls or coco-coir soil alternatives and flooded or submerged in water. Nutrients for hydroponics are provided in solution in the water. For aeroponics, the roots dangle directly in the air and the nutrient salts are mixed with water and sprayed as a vapor directly onto the roots. This completely eliminates mechanical resistance. Roots can grow and expand their surface area at will.

## Oxygen Impact

Oxygen and moisture are key to the process of nutrient absorption in the cycle of photosynthesis. Nutrient salts move through the plant root surface along with water and oxygen to begin the conversion cycle. These nutrients transport up into the canopy level as compounds where photosynthesis uses light energy converting $CO_2$ and nutrient salts into plant growth, while releasing oxygen and water into the air. So the summary is roots need oxygen while the canopy needs carbon dioxide.

The difference in aeroponics vs. hydroponics vs. soil beyond the surplus of oxygen is control. Soil can be very forgiving. Plants in soil will grow naturally at a slow rate as roots extend their way through the soil. Soil can stay moist for extended periods of time as plants grow. Hydroponics and aeroponics accelerate this process by providing nutrients directly to the roots. Hydroponics has the downside that oxygen levels have to be managed over time. Standing water, depleted oxygen, pH levels, and nutrients can trigger algae growth and fungal problems in hydroponics. This requires steady attention. Hydroponic nutrient dosing is normally managed on a batch volumetric level. Everything is mixed at one time, irrigated, and when it runs out, mixed again. Managing nutrient dosage, water acidity levels (pH), and oxygen mix is a complex process in-solution. Providing the same nutrient levels to hydroponic plants across this cycle and a large irrigation area can be a challenge.

Aeroponics supercharges plant growth with a surplus of oxygen at the root surface. When this is combined with sensor technology and dynamic nutrient recipe dosing, plant growth is superior.

Aeroponic nutrient "dosing" can be precision optimized for nutrient dosage, spray time, light synchronization, growth phases, flowering, bloom, and fruiting cycles and pH levels, to maximize results. In a proper clean room environment, aeroponics delivers pure fresh results, with zero pesticides, highest quality flavors, and maximum growth.

Precision Sensors, Advanced Automation Software, & Nutrient Recipes Offer Growth Breakthroughs With Aeroponics!

- Rapid Aeroponic Growth = Yield & Profits
- No Pesticides, No Pesticide Residue = Higher Quality
- Pure Water: No Heavy Metals No Pathogens, No Antibiotics, No Hormones = Just Great Taste!
- Automation Simplifies The Process - No Disruption, More Growing Time = Lower Cost
- Reduce Stress - Plants Start Healthier = No Mechanical Resistance From Soil & Create Greater Yields Faster
- Internal Methods Protected - Proprietary Nutrient Recipes, Irrigation Cycles, & Light Automation = Proprietary Advantage

## Automation, Control, & Sustainability

Fresh Taste, Perfect Nutrient Recipe, Every time!

Aeroponics takes the best of hydroponics and expands it further with greater water savings, energy efficiency, and ample root zone control for accelerated growth.

Using the latest precision sensing technology and software controlled dynamic nutrient dosing, growers are given an unprecedented level of environmental control for their growing needs.

Nutrient levels delivered in leaf, bloom, or fruit tie directly to the nutrient recipe and the ability of the roots to absorb the nutrients converted through the photosynthesis process.

## Root Zone Control

More Oxygen & No Mechanical Resistance from Soil

Unlike standard hydroponic systems where plant roots are typically submerged in water, aeroponic roots hang in the open air with no mechanical resistance from soil. This enables the roots to grow with abandon to support much larger foliage, bloom, & fruit growth in the canopy.

Dangling roots not only absorb essential minerals from the nutrient spray solution, but this also allows increased oxygen intake to fuel respiration.

This accelerates growth +40% more than in soil, and adds the benefit of greatly reducing nutrient usage (-70%) & water usage (-90%) through scheduled spray-recycling.

**Keep Your Facility Pure & Pest Free**

Using The AEtrium System Eliminates Soil Threats

Growing in a scientifically controlled clean rooom environment seals out pests. Remote access controls the automation of the growing cycle with the Guardian™ Grow Manager Software combined with reliable mechanical grow environments (AEtrium-2.1 SmartFarm and AEtrium-4).

Add a positive pressure air flow clean room environment combined with air lock access and you shut out the source of 99% of pest infestations (people). Manage your grow room environment like a modern surgeon. Use automation to increase productivity and deliver superior results.

# BASIC HYDROPONIC EQUIPMENT

## CONSIDERATIONS FOR PURCHASING YOUR HYDROPONICS EQUIPMENT

Entering the world of hydroponics is very fun and fulfilling, but it can also be somewhat intimidating. There are so many options for purchasing your hydroponics equipment that it can feel overwhelming. While it can be tempting to purchase the first system that looks cheap and easy to use to you, you don't want to wind up with equipment that doesn't match your needs. Here are the most important factors consider when buying your equipment.

**Your Available Space** – Where exactly will you be growing your plants? A small greenhouse in your backyard? A large closet? Your basement? Before you purchase your equipment, make certain that you calculate the square footage of the space you will be using and figure out exactly how much hydroponics equipment you can put in there. If you are planning on growing rows of plants, try to allow at least one meter of walking space between each row to make it easier to tend to your garden.

**Your Plants** – You probably already have an idea of what you want to grow hydroponically. Now you need to make sure that you find hydroponics equipment that can help those plants grow their potential. You certainly don't want to purchase small, shallow trays if your plants have larger, thick roots. And you don't want to waste your money on several eighteen-inch buckets if all you want to grow is smaller plants. Talk to your hydroponics retailer about what kind of system, medium, and fertilizer would best accommodate the size and growing rate of your plants. Many manufacturers also have phone numbers that allow you to talk to hydroponics professionals about these kinds of growing issues.

**Your Budget** – Before you purchase your hydroponics equipment, you should decide upon how much money you are willing to spend, and try to make the absolute best use of that budget. It is important to keep in mind, however, that start up costs aren't the only expense related to hydroponics. You should also try to factor in how much energy your lights will require and how often you may have to replace your equipment. If you plan on keeping your hydroponics system for years, it can save you a lot of money to spend a little extra when you actually buy the equipment.

**Your Time** – Like most hobby growers, you probably don't want to devote all of your time to growing you plants. This is why you should also consider exactly how labor-intensive individual systems are. Something like an aeroponics system might seem immediately appealing. But since anything that goes wrong with the timer would result in a very quick drying out of the roots, these kinds of systems can sometimes require more attention than most. Most people simply don't have the luxury of rushing from work to their home to save their plants in the event of a power outage. So look for a system that provides you with a larger margin of error, such as one that accommodates a medium that holds a great deal of air and water well.

Hydroponics can at times seem overly complex and certainly overwhelming. The various styles and methods of hydroponic gardening only further exasperate the situation.

There are however basic hydroponic equipment and supplies that are universal regardless of what mode of hydroponic gardening you choose.

Properly executed hydroponic growing can result in phenomenally higher yields as opposed to soil based gardening. Cost effective use of space and time leads to a more rapidly maturing crop with higher yields.

- Reservoirs
- Water Pumps
- Timers
- Grow Media
- Light
- Air Pumps
- Nutrients

## Hydroponic Reservoirs

A reservoir is fairly universal in hydroponic setups. It is simply a container where your water - nutrient mix is stored and drawn from. The fundamental theory of soil-less gardening aka hydroponics is to keep your plants root systems awash in a balanced mix of water, nutrients and air. The reservoir stores the water, and the water stores the nutrients.

Reservoirs can be anything from elaborate and expensive commercial versions, down to a simple five-gallon bucket. The larger ones are naturally better, and for a midsize to full scale home hydroponic operation they are generally more cost effective.

Any reservoir you decide to use must have a lid or you will be losing fluids to evaporation faster than you would ever think possible.

The loss of fluids is not simply a lower fluid level - it also louses up your nutrient ratio and contributes to salt and mineral build up which can become a real problem.

It is also crucial that the reservoir is not metallic, any metals are more than likely to introduce harmful mineral elements and instigate chemical interactions that can be damaging and deadly to your plants.

**Water Pumps** ~ Delivery System

Water is the key to all life as we know it, plants are no exception. Irrigation is obviously necessary for any type of gardening, more so in hydroponics as the soil has been replaced with a media and fluids. A water pump to circulate water and nutrients your plants makes the entire set up viable.

Water pumps are measured by either "GPM" which is gallons per minute or "GPH" which is gallons per hour. Larger commercial pumps can circulate thousands of gallons per hour. Small output pumps which are fairly inexpensive put out as little as 30 - 40 GPH, which is all that is really necessary for smaller setups.

The substrate/grow media you select will also have an effect on the irrigation equation. Large, smooth rounded growing medium drain quicker and generally need frequent watering cycles, while moisture retentive porous media drain more slowly.

**Timers**

With a water pump, a timer is also advisable. You could manually turn the pump on and off at the correct intervals, but a timer is more efficient.

A timer shuts off the pump at pre-set intervals and allows nutrient solutions to drain back into the reservoir awaiting the next feed cycle. For most plants and setups, as little as 4 cycles daily is sufficient, some plants and setups perform better with about a dozen daily cycles. If you are uncertain what cycle to use for your setup and crop - choose the middle road to start with, 6 -8 cycles daily and adjust your schedule from there.

**Air Pumps**

All plants need air and in particular the oxygen and carbon dioxide ($CO_2$) it is comprised of. Some hydroponic systems are somewhat self-oxygenating but none are truly and wholly self-aerated. An airstone / airpump combo is not an absolute necessity for a hydroponic system, but it is advisable.

Some Deep Water Culture (DWC) hydroponic systems incorporate what is known as a bubbler. Net Pots sit in nests in the top of the reservoir while the bubbler basically carbonates the water and provides ample air and oxygen.

The simplest air pumps are the same as those that are used in aquariums, a small 5 watt pump with an airstone attached to a tube.

**Hydroponics Nutrients**

An essential element to your list of hydroponics supplies is nutrients. Nutrient is the fertilizer and plant food. Nitrogen, Phosphorus, Potassium and trace elements in a water soluble format. Requirements vary from plant to plant the same as they would in nature.

Any nutrient formula will supply the 3 key elements Nitrogen, Potassium and Phosphorous. However, a problem sometimes associated with hydroponic gardening is that plants do not always have access to the same trace elements that they would normally find in the soil. That doesn't mean you should go out and spend exorbitant capital on designer nutrients. Shop for hydroponics nutrients that provide trace elements, secondary and micro-nutrients. These will be listed on the label.

## Hydroponic Lighting

Plants require both light and dark to survive and thrive.

Plants take in carbon dioxide during the daylight while involved in photosynthesis. At night in the darkness, they release carbon dioxide. It's the equivalent of breathing, you can't continuously inhale and never exhale - it's a physical impossibility. The same is true with plants they need a separation of day and night, light and dark cycles.

Grow Lights, Hydroponic and Horticultural lighting, allows you to extend the natural growing season of plants by providing your plants with artificial sunlight.

## Growth Media ~ Substrates

Hydroponic gardening is soil-less gardening. No soil! However, the plants need to be supported or held up. This is done with Hydroponic Growing Mediums which is a soil-less media... inert, non-organic materials. Hydroponic Growing Mediums act as an anchor to the plant to prevent it from falling over as it grows, provides good drainage of nutrient solution and allows the flow of oxygen to the plant roots.

Many things will suffice as growing media, some perform better than others under varying situations. They should all meet the following very basic requirements.

1. A Hydroponic growing medium should be reasonably dense in order to provide enough weight and mass to anchor a plant, but not so dense that it will impede the flow of oxygen and nutrient solution.

2. It must be clean and/or sterile to impede the spread of disease, pest and parasitic organisms.

3. Each individual growing medium 'nugget' must be small enough to provide a larger amount of surface area to remain damp with nutrient solution between flooding allowing the plant to feed.

# COMPARING HYDROPONIC WATERING SYSTEMS

A Common Choice: Hydroponic Drip System

There are different places where hydroponic system plans can be found, either for free (DIY) or for purchase at a retailer. There are companies that will sell hydroponic drip systems pre-made. These systems can be more simple than trying to follow a plan to build a drip system. Building a system from scratch, although fairly simple, still leaves the possibility of not doing something correctly which will make the system ineffective. Buying a pre-made system allows you to have a system that is up and running quickly and that is guaranteed to work for years to come.

**WICK VERSUS DRIP**

There are two systems that are most commonly made from scratch and those are the wick system and the drip hydroponic system. The wick system is very simply constructed. It uses a tray that holds the plants which sits on top of the reservoir that holds the water and nutrient solution. These two pieces are held together by a wick which allows the water solution to transmit these elements into the tray with the plants. The down side to this system is that there is not a way to regulate the amount of solution that is getting to the plants, which means that a small amount of plants can be grown at one time. This would make an ideal choice for someone wanting to grow a limited amount of plants or with very limited space.

The drip hydroponic system is built in much the same way with the tray that sits on top of the reservoir. In this system, however, instead of using wicks to transmit the solution to the plants, there is a submerged pump system that transmits the solution to the plants at whatever rate the gardener sets on the meter. The solution goes into the tray a drip at a time so that there is a constant flow of nutrients without the plants becoming overly submerged like in other systems.

There are benefits to using the drip hydroponic system, such as the fact that you do not have to monitor the system as often as other systems because the pump and timer handles everything automatically. Also, since the flow rate can be regulated, plants tend to grow larger, have a higher yield, and higher numbers of plants can be grown at one time in the system. One down side to using the drip hydroponic system is that it is more expensive than the wick system since the pump for the system has to be bought as well. However, this is a popular system, especially for use in greenhouses where larger numbers of plants are grown at a time.

# POPULAR HYDROPONIC KITS

As hydroponic gardening is becoming more popular, the sales of hydroponic kits has become a booming business. You can purchase a kit to make a small garden or one to build an entire hydroponic room. The items that are included in these hydroponic kits will depend on the type of system that you wish to use.

**The Aggregate Hydroponic Kit**

Hydroponic kits for an aggregate system will usually begin with a container that is not transparent in which to place your plants. The company will also include a separate nutrient tank with the nutrient solution included in the kit. There will be some type of hookup system that will allow you to run the solution from the nutrient tank to the plant reservoir in order to perform the flooding process that is necessary to keep the plants fed. A built in drainage system is also a necessity with these hydroponic kits.

Aggregate hydroponic kits will naturally include a type of aggregate (the substance that you will place your plants in to give them the support that they will need). A pH tester might also be included to ensure that your nutrient solution is at the proper pH level.

**The Water Culture Hydroponic Kit**

Hydroponic kits for a water culture system can be simple or complex, depending upon the amount of money that you wish to spend. A simple water culture hydroponic kit can include a small plant container, a simple lighting system, and a nutrient solution. The more complex models can contain an air system, a water heater, and various testing materials for your nutrient solution.

## The Aeroponic Hydroponic Kit

Hydroponic kits that use the aeroponics system are the most advanced and often the most expensive. One brand of these kits comes with pots that have plants already in them as well as a plant holder. It also includes a submersible pump and tubing as well as three separate spray nozzles. Also included are enough liquid nutrients to make one hundred gallons of nutrient solution and a pH test kit.

## Additional Items that You May Need for Your Hydroponic Garden

There are a number of other items that you will need for your hydroponic garden that may not be included in the hydroponic kits. You will need to purchase a lighting system (if one has not been included). You should purchase lights are large enough and have enough wattage for the size of the area in which you are using them. You may want to install blue lighting if you are dealing with less mature plants. The more the light that your plants have, the healthier they will be.

Other items that are not included in most hydroponic kits are measuring devices, such as a thermometer and a humidity gage. The latter is especially vital when using an aeroponic system. A thermometer can be used to make sure that the area where you have your hydroponic garden is warm enough to sustain your plants. You need to make sure that your plants are kept warm, especially if your garden is located in a basement during the winter months.

## Hydroponic Kits - Great for Beginners

Hydroponic kits are a great way to build a hydroponic garden if you are a beginner and are not sure about how to get started. Their clear, simple instructions and basic supplies can help you on your way to becoming a successful hydroponic gardener.

# AEROPONICS SYSTEMS

Aeroponics systems use 100% humidified AIR, instead of soil or water, as the growing medium to cultivate plant life. The plants roots dangle into a growing chamber which is kept at 100% humidity by misting nozzles, which spray a misted nutrient solution at periodic intervals.

This allows the plant to absorb on the nutrients and moisture it needs Since their inception some 30 years ago, it is fair to say that aeroponic techniques have proved very successful for propagation, but have yet to prove themselves on a commercial scale. They are very popular for indoor herb and vegetable gardens. They been tested, proven and refined for many years and are used by scientists around the world, including NASA. Aeroponics systems are the smart choice for year round gardens indoors or out. Grow plants without soil and achieve a greater bounty of fresh fruits, flowers, vegetables, and herbs over conventional gardening methods. Aeroponics systems do not suffer many of the problems of traditional gardens such as, root rot, weather damage, drought, losing produce to predators, soil degradation, and others.

**Advantages**

The Plants roots are also able to access as much oxygen or $CO_2$ (carbon dioxide) as they need. This increases the efficiency of photosynthesis and the growth cycle of the plant can also be increased dramatically using aeroponics systems instead of other available growing mediums.

A further advantage is their ability to ease the propagation of plants from cuttings, due to the lower bacterial and pathogen levels. This enables cloning of your favorite plant varieties. Because they conserve water and energy, often using 1/10th the water of conventional hydroponics systems.

Universally considered to be an ecological and economically friendly method for producing healthy, natural plants and crops, Aeroponics systems are fun, simple and easy to maintain. They provide an effective and automated method to provide your plants optimal levels of water, nutrients and oxygen for rapid growth and higher yields. Experience faster growth rates, higher yields, increased nutritional value, heightened fragrances, and ease of maintenance with fully automated aeroponics systems.

**Disadvantages**

Aeroponics systems do have relatively high setup costs and are also quite mechanically complicated and susceptible to malfunctions. These malfunctions of the mechanisms used to control the precise regulation of water and nutrients can damage plants in a very short amount of time. Therefore, regularly monitoring for blocked nozzles or breakdowns is necessary. (End disadvantages)

Advances in technology and growing popularity of aeroponics systems moving forward will continue to improve the results gained via aeroponic gardening practices. Aeroponics offers the chance to vastly improve the efficiency and effectiveness of not only our home gardening, but also food production on a commercial scale. I firmly believe that aeroponics will continue to grow in popularity and become a staple technology within the plant and food industry throughout the 21st century. Developing countries and countries with limited natural resources, especially water, will have no choice but to implement aeroponics systems technologies and the rest of the developed world will definitely take notice of the quality, quantity and environmental friendliness of the plants and foods produced.

## BASICS AND BENEFITS OF AEROPONICS

If you are a new indoor gardener or a long-time horticulturalist, you will find that aeroponics is a type of indoor, soilless gardening similar to hydroponics that is gaining in popularity. Aeroponics systems allow gardeners to grow plants where the soil is poor or nonexistent, and the plants grow in air rather than soil. For those in the nursery trade, an aeroponic or aeroponic cloner can help you quickly start cuttings from all manner of established plants, which can then be transplanted and grown on for resale. For their convenience of use and speed in producing healthy and large plants, gardeners turn to aeroponics systems for their indoor gardening setup.

To get started with soilless indoor gardening, you might want to start with hydroponic grow kits. Hydroponic grow kits offer all of the pieces and parts of a complete growing system so that you can easily begin the process of growing your own food or flowers indoors. Hydroponics uses a soilless growing medium in which the roots of the plant are grown. It might be coir or other types of medium, and a reservoir system is put in place under the medium holding the roots. In this reservoir system is placed a solution of water and nutrient, which is a soluble plant food that hydroponically grown plants require for food.

An allows the roots of the plants to be grown in the air. This is accomplished by suspending the crown of the plant in some kind of a holder, such as a net, which is placed in an open-bottomed tray. In this way, the roots are exposed to the air. Normally, a plant would receive water and food through the roots via the soil, but in an aeroponic system a misting device is used to spray water and nutrient onto the plants' roots. Because the roots of plants in an aeroponic system are suspended in the air, the roots are exposed to a great deal of oxygen, and this is one big reason why aeroponically grown plants grow very quickly and reach substantial sizes.

There are several quality brands of aeroponics systems available. Lines such as AeroFlo, American Agritech and Rainforest all offer top quality aeroponics systems suitable for home or commercial gardeners. These systems are flexible enough to allow you to grow dozens of plants in a small area, and they create lush foliage, flowers, fruits and vegetables in a short amount of time. You will find an aeroponic system, hydroponic grow kits and an aeroponic cloner at better online discount gardening supplier websites.

Man continues to discover new ways of growing their plant foods. One of the latest techniques is growing with aeroponic gardening machines, which is basically a hybrid of the hydroponics method of gardening in that the plants are suspended in air with the roots kept as moist as possible through a mist of nutrient solution.

## Benefits of the Method

As the years go by, it seems that conventional farming is becoming more of a burden than a blessing. For one thing, huge amount of synthetic pesticides and nutrients are washed away to natural bodies of waters (rivers and lakes) whenever it rains hard, adding up to the pollution caused by industrial manufacturers. What is more, the astronomical amount of chemicals used by the agriculture industry significantly contributes to the increasing amount of greenhouse gases in the atmosphere.

Anticipating such situation, scientists have been exploring innovative farming methods that pose no harm to the environment. One particular promising technique that they have come up with is vertical farming. With this method, farms are set up in high-rise buildings and use no soil. Water is used efficiently and nutrients are applied to the plants in precise ways.

An aeroponic system is one of those techniques used in vertical farming. Because of advances in technology, ordinary home gardeners today can avail of this technique. This may be a new gardening technique and may still be initially costly to avail of. However, there are huge benefits aeroponics can present to any avid home gardener.

**Healthy Produce**

You might have heard of the EU cucumber scandal where organically produced cucumbers got infected with the deadly E. Coli bacteria. The possibility of picking up an infected vegetable at the grocers is really a frightening prospect. For all of us value the health and well-being of our loved ones.

So how can you make sure that your family will not eat contaminated vegetables? The best way is to produce your own vegetables. By utilizing an aeroponic system of home gardening, you will be assured that your vegetables will be free from harmful bacteria. Most of the time bacteria get into the plant via the soil. Because aeroponics does not use any soil, there is no chance for harmful microbes to infect your plants.

## Ideal for Limited Space

One of the biggest problems urban gardeners face is the insufficient outdoor space. But why limit your potential production when you can increase it by using aeroponics? Aeroponic system does not require any substrate. You can, therefore, maximize available space by arranging your plants vertically.

## Efficient

One of the biggest advantages of aeroponics is that it is very efficient in utilizing resources. You will not spend that much water with aeroponics as you would with conventional gardening. Only required volume of water is applied by controlled misting systems. Also, there is no chance for wasting nutrients. You just put in the required amount of nutrient along with the water and the nutrient-rich concentration will be sprayed by the misters with no run-off whatsoever.

Aeroponics gardening has been used to grow fresh vegetables in space as well as in large-scale farms. However, many manufacturers now offer home aeroponics kits to gardeners, which make it easier to learn the basics of the method.

And then there is the fact that aeroponics gardening is more economical than hydroponics. This is because the former does not use as much water and nutrient solution than the latter since the plants are just misted with them instead of being submerged in them. For this reason, too, aeroponics gardening promotes better plant growth virtually free of diseases, thanks to more exposure to oxygen.

Since the aeroponics garden takes up very minimal space, even city dwellers living in cramped apartments can engage in it. The trick is in making sure that all the other growing factors are present in the area selected.

**Basic Components**

When you buy ready-made aeroponics kits, you will be getting basic components such as the growing bulbs, the basin/container for the nutrient solution and water, the growing platform, the stand with light hood, and plant pods. Of course, instructions are also provided for the installation and maintenance of the aeroponics set-up.

You have to read said instructions manual as your success depends on how well you understood the tips contained therein. You should only take about 30 minutes to set up the whole thing especially with the most basic models.

## Basic Things to Remember

Before setting up your aeroponics kit, you must identify the best location for it inside the house. There are many ideal locations such as the shelf, the desk and any water-resistant surface. We do not recommend placing it on your hanging kitchen cabinets as these are too close to the counters.

Indeed, many aeroponics kits require a partially or totally closed environment for many reasons. First, the entry of pests and bad bacteria can be avoided. Second, curious household pets have lesser opportunities to make a mess of the system. Third, it is better to contain the nutrient-rich misting solution in the air if the environment is closed off from the outside.

If you intend to go into aeroponics cloning, you will need a chamber or reservoir.

This component will provide the proper environment in which cultivation can occur. You will also need a pump to evenly distribute the water and nutrient solution to the plants as well as a timer to ensure consistent application.

Aeroponics gardening is a possible thing to do in the comforts of your own home.

No knowledge in rocket science is necessary, no complicated equipment is required and no great deal of money is needed. You just have to exercise patience and perseverance with a dash of gardening passion to learn the basics of this new plant cultivation technology.

**SUCCESSFUL GARDEN INDOORS**

You might have wanted to garden indoors, but have been stopped short because you have no suitable area in which to garden. There are ways, however, which everyone, even those who live in cities and have no backyard, or those at the tops of apartment buildings, or those with inhospitable climates, can utilize in order to grow plants indoors. One of those methods uses an aeroponics system for growing plants indoors. An aeroponic system allows the gardener to grow plants without using soil, a great benefit if you have no soil or it is inconvenient or impossible to get soil to your home. Instead, an aeroponics system allows the plants to grow in the air, with the roots exposed to air.

Although you might have heard of an aeroponic hydroponic system, an aeroponic hydroponic system is really a misnomer. While plants grown in an aeroponic system grow in the air without soil, hydroponically grown plants also do not use soil, but they do instead use a soilless growing medium, such as coir, in which the plants are grown.

One of the best ways to get started with setting up your own aeroponics system is to use aeroponic kits. Aeroponic kits have all of the individual components that you will need for this form of gardening. Aeroponics kits will contain some means of keeping the crown of the plant supported while the roots dangle freely below. Special types of foam or even netting is used in aeroponics kits to suspend the plants in the air. Like any plant, those grown in an aeroponics system require food. The food is known as nutrient, and it is dissolved in water to form a nutrient solution. The nutrient solution reaches the plant roots through the use of aeroponic misters, which gently spray the roots of the plant with nutrient at regular intervals.

There are benefits to using an aeroponics system. One is that the roots of each plant is exposed to a great deal of oxygen. This is important, because plants need oxygen so that they can grow strong and reach maturity rapidly. Another is that fewer plant diseases cause trouble in aeroponics, because the plant is never exposed to soil-borne disease.

## DIFFERENCE BETWEEN AEROPONICS AND HYDROPONICS

With the agricultural breakthrough that has come over the past century, planting has become easier than it was before. Two of the breakthroughs, the hydroponics and aeroponics have come a long way in developing the agriculture and food production that not only resulted to individual progress but also the economic growth. But what is the difference between aeroponics and hydroponics?

The difference between aeroponics and hydroponics is that hydroponics is the means of soilless planting where there is a little or no soil required at all. The nutrients are supplied through a nutrient solution or a nutrient film at which the formulation has been controlled to sufficiently meet the needs of the plants. Aeroponics meanwhile is similar to hydroponics, growing plants without the use of the soil. The difference between the two is that the latter uses nothing as medium while the former uses water.

However, many has become confused of the agricultural methods and the difference between aeroponics and hydroponics. Aeroponics is a form of hydroponics procedure. It is the water that serves as the nutrient carrier when then the nutrient solution is sprayed into the roots of the plant.

Another difference between aeroponics and hydroponics is that the environment set-up. While in hydroponics, the plants are cultured into an enclosed area like the greenhouse, the plants grown through aeroponics is in a closed or semi-closed area. The environment is not that restricted which may become a source of potential problem. This is for the reason that the plants grown aeroponically are not kept free from plants hazards like pests and diseases.

One significant difference between aeroponics and hydroponics is that plants can mature easily with air as there are abundant sources of the essentials in plants' growth. These are oxygen, the nutrient solution and the water.

You might quetsion why there are some farmers that prefer aeroponics rather than hydroponics. This is for the reason that aeroponics can give consistent supply of oxygen that can yield to a much higher yield of production.

The procedure between hydroponics and aeroponics also differ in terms of set-up. The aeroponics procedure allows the plants root to be suspended in a hydro-atomized nutrient solution that will make some parts of the roots like the crown to be extending on the top. In comparison to hydroponics, the procedure varies depending on the technology used. These technologies ranges from static solution and continuous flow solution. Moreover, there are also some precautions made to avoid disease contamination in the irrigation system.

There are a number of difference between aeroponics and hydroponics. Despite of these, it cannot be denied how these technologies improved the food production and the lives of many.

## A DEEPER LOOK

As our soil quality begins to deteriorate, many people are looking for alternative methods to grow fresh vegetables for their future homes. Because of this reason, hydroponics agriculture will become more popular in the future.

Aeroponics is a simple concept yet it is the most technical among all 6 types of hydroponic systems. Aeroponic system is used by a lot of home growers because it has brought really good results for them.

Let's get into details about this technique.

Aeroponics is actually a subset of the hydroponic system. However, with the hydroponics method, plants use water as the growing medium while aeroponics uses no growing medium at all. This technique was invented during the 1940s and since then, many researchers have added to the theory and application of this method. Aeroponics is considered one of the best methods to grow plants in a soil-free environment and the need for this method has been growing due to a clear need for a more convenient way to grow plants.

In the aeroponic system, plants are not contained in any solid material such as Rockwool or soil. Instead, plant roots are hung in the air in a grow chamber in a closed-loop system. The roots are sprayed with nutrient-rich water or fine, high-pressure mist containing nutrient-rich solutions at certain intervals.

This makes aeroponics a more advanced form than the hydroponic wicking systems, deep water culture, and other types

As all plants need nutrients, the organism will spend a valuable amount of energy growing roots to find these pockets of nutrients in the soil for flower formulation and growth. With hydroponics and aeroponics, nutrients are instead delivered straight to the roots.

Compared to regular hydroponic plants, the plants tend to grow faster and absorb more nutrients because the roots are exposed to more oxygen. Also, there are fewer threats of diseases around root zone disease because there's no place for debris or pathogen to reside.

However, you need to be aware of the fact that aeroponic system chambers are constantly wet with nutrients spray which is convenient for harmful bacteria and fungi to develop. Therefore, it is important to clean and sterilize these misters before using, and occasionally take out and keep these chambers treated with the hydrogen peroxide solution, which can be purchased at any quality hydroponic store.

## HOW DOES THE AEROPONIC SYSTEM WORK?

In the aeroponic system, plants are usually inserted into the platform top holes on top of a reservoir and placed within a sealed container.

Due to no root zone media for plants to anchor in, you need to prepare a support collar that will hold stems in place. These collars must be rigid enough to hold plants upright and keep the roots in place but flexible enough to allow room for roots to grow.

The pump and sprinkler system creates vapor (which is a hydro-atomized spray mixture of water, nutrients and growth hormones) out of the nutrient-rich solution and sprays the mist in the reservoir, engulfing the dangling plant roots and absorbed by them. This spray provides the exact amount of moisture which stimulates the plant's growth and allows it to develop turgidly.

The timer supplies the timed spray intervals and duration for the plants. Some people think that growing plants in aeroponic system would be frailer compared to hydroponics. But that's not true. The secret of aeroponic system all lies in the amount of oxygen exposed to the roots without a root zone media limiting it.

Thanks to this, the plant roots will develop rapidly and grow in a moist air-rich environment. If you want to see their development rate, just lift up the growing chamber to see how they are growing.

## Types of Aeroponic Systems

Low-pressure Aeroponics (LPA)

This is the most commonly used aeroponic type used by most hydroponic hobbyists due to its ease to set up, availability at any hydroponic shop, and low cost.

Low-pressure creates droplet size much different from the high-pressure aeroponic system.

What you need for this system is just like any hydroponic system – a pump strong enough to move the water onto the sprinkler heads to spray water around the plant root zone.

### High-pressure Aeroponics (HPA)

This type of Aeroponics is more advanced and quite costly to set up as it would require specialized equipment. So they are often used in the commercial production rather than home growers.

The HPA must run at very high pressure to atomize water into tiny water droplets of 50 microns or less.

This system creates such a fine droplet size that create more oxygen for the root zone than the LPA, making it the most efficient system among all aeroponic types.

### Ultrasonic fogger Aeroponics

Ultrasonic fogger Aeroponics, or commonly called fogponics, is another interesting type of Aeroponic system.

As the name means, growers would use an ultrasonic fogger to atomize water into super small droplets of water. These are very tiny and you will see it in the form of fog.

Though plants roots find it easier to absorb water in tiny size, there's little moisture in the fog created, and when running over time, it can easier create the salt that can clog these foggers than other Aeroponic types.

### WHICH PLANTS TO GROW

You can use this system to grow nearly any type of plants and cultivars such as vegetables, nursery stock, houseplants, and bedding. Hundreds of species of plants have been tested and grown successfully by commercial greenhouse owners, researchers and nursery operate using this technique.

**TOOLS NEEDED**

What you will need:

• A reservoir/container to hold the nutrient solution

• Nutrient pump

• Mist nozzles

• Tubing to distribute water from the nutrient pump to the mister heads in the growing chamber

• Baskets to suspend plants

• Enclosed growing chamber for the root zone

• Watertight containers for the growing chamber where the plant's root systems will be

• Timer (preferable a cycle timer) to turn on and off the pump

You can purchase these tools at the local gardening/hydroponic supply store or online.

It's quite easy to understand how an aeroponic system works. The purpose of the plant roots hanging in mid-air is to get them exposed to oxygen as much as possible. The high volume of oxygen exposure will stimulate their growth and help them grow faster than they would in the soil, which is a very important benefit of this type of system. This can be seen in the massive root growth of the plants. It has also been proven that this technique increase crop yield X10 compared to using soil. Not only will you be able to collect healthier plants but also grow more crops per year.

No growing media is needed with aeroponics. You can use baskets or closed cell foam plugs (compress around the plant's stem) to suspend the plants. These tools fit in the small holes on top of the growing chamber. In the growing chamber, the mist nozzles will spray the nutrient solution to the plant roots at short intervals. This regular spray has benefits of keeping the roots moist and provide the nutrients they need.

The growing chambers should be airtight and light proof (to prevent the root zone being penetrated by lights and make algae cannot thrive). It must allow air to get in for the growth of plant roots but you also don't want pests to get in or water to spill out. The chamber needs to be able to hold in the humidity as well. The ultimate factor which creates a successful aeroponic system is a balance of plenty of moisture, nutrients and fresh oxygen that you can provide to the roots.

Finally, a major factor which contributes to the success of this system is the water droplet size. A fine mist would create much faster-growing and bushier roots. These roots also have more surface area to absorb the oxygen and nutrients compared to those sprayed with small streams of water from small mist nozzles. This will also mean faster-growing plant canopy.

## PROS AND CONS

PROS

Benefits of aeroponics include:

• Maximum nutrient absorption for plant roots due to no growing medium

• Massive plant growth because plant roots are exposed to oxygen 24/7. This promotes healthy and fast-growing plants. The mist used on the roots can also be sterilized to prevent plant diseases.

• Higher yields

• Considerably fewer nutrients and water used on average compared to other systems because of higher nutrient absorption rate. You can help the environment by using less water and human labor

• Mobility. You can move around easily plants or even the whole nurseries as all you need is to move the plant from one collar to another

• Little space required. You don't need a lot of space to be able to set up this system. Plants can be added up one on top of each other. With this type of modular system, you can maximize the use of limited space.

• Easy system maintenance because there is no growing medium used. However, you need to disinfect the root chamber regularly, and periodically the irrigation channels and the reservoir

• System can be cleaned easily

• Easy to replace old plants with new ones

• Great educational value. Adults and kids can use this system to grow pet plants and learn a great deal about plants without needing to get their hands dirty.

**CONS**

• Besides many great advantages, aeroponics also has the downsides that cannot be overlooked such as:

• Require constant attention with pH and nutrient density ratio because this system is sensitive. Understanding what is the right ratio and applying this concept may be difficult for beginners and can only be attempted by those who are more familiar with such systems.

• The cost for initial set up can be high, which can be many hundreds of dollars each

• Require constant supervision

• Susceptible to power outages. You will have to water your plants manually if this happens.

• Require technical knowledge as it one of the most technical to set up. As there is no soil to absorb the excess nutrients, you need to have a good knowledge about amounts of nutrients required by the plant roots

• Dependence on the system. An aeroponic system is made of mist nozzles, high-pressure pumps, and timer. If one of these breaks down, your plants will die easily.

• Require regular disinfection of the root chamber (a common disinfectant is hydrogen peroxide) to prevent root diseases.

• Microorganism can be introduced to plant roots through water.

The best benefits of aeroponics are the massive plant's growth and higher yields compared to other systems. However, these advantages also come with a cost. The price to set up the system is quite expensive and it also requires technical expertise as well as advanced knowledge (about pH and nutrient density ratio) for this special kind of plant cultivation. If you are a beginner in the domain of growing hydroponically, I would say really weigh the pros and cons and choose the best system that you think will work for you at this stage.

# DEEP WATER CULTURE

Deep water culture hydroponics is a method of growing plants where the roots are suspended in an actively oxygenated solution of water and nutrients, rather than planted in soil. It eliminates many of the drawbacks of growing plants in soil and results in faster plant growth and larger crop yields.

There are many different ways to build a deep water culture hydroponics system, each with advantages and disadvantages for the user, depending on their goals. The simplicity of this agriculture technique is that by using a suitable well-oxygenated water solution, plants can be grown efficiently, with significantly reduced labor requirements than soil based cultivation.

## An Introduction To Deep Water Culture Hydroponics

Growing herbs, flowers and other plants at home is becoming increasingly popular. For those who want to try something a little different than growing plants in soil, hydroponics is a great option. Deep water culture systems are a great way to get started with hydroponics. This growing method can be as simple or as complicated as you want it to be and can be scaled up as you gain experience.

Deep water culture hydroponics is not only popular among homeowners, but many commercial organizations use deep water culture systems to grow a variety of fruit and vegetables crops in an efficient and sustainable way.

Deep water culture hydroponics provides an efficient and sustainable way to grow a large variety of plants that can be used for food in the kitchen, or for commercial production by food producers.

If you are new to the technique of growing plants using deep water culture hydroponics, then this guide is perfect for you. We will explore what deep water culture hydroponics is all about, take a look at how the process works, and guide you in the direction of creating your own system, which you can use to grow plants at home.

## Components Of Deep Water Culture Hydroponics

The solution in which the plants grow needs to be enriched with certain components in order to make sure it provides plants with all the necessary elements for their growth and nutrition. These components can be highlighted as follows:

- Oxygen: Since plants are living organisms, they require oxygen for their respiration. Roots of plants in soil receive oxygen through the gaps present within the soil particles. In a deep water culture system however, oxygen needs to be pumped into the water solution to make it available for the root cells.

• Water: Essential for a plant to grow and thrive, water is abundant in a deep water culture system. The challenge comes in ensuring that the large volume of water in the system does not cause problems in the delivery of oxygen and nutrients to the plants.

• Nutrients: The water in deep water culture systems needs to be supplemented with nutrients required for plant growth, which would otherwise be absorbed via the soil. Adding the correct amounts of nutrients into the water solution makes sure the plants are well-nourished and able to develop to their full potential.

### Advantages of Deep Water Culture Systems

Deep water culture systems are a good way to get started with hydroponics. It is one step up in complexity from wick systems, but it is still simple enough to be accessible to anyone. Here are some of the main advantages of deep water culture systems.

• Simplicity of setup. Deep water culture systems are easy to set up and only require a few parts that can be put together in a short period of time. The only moving part is an air pump, which is easy to configure.

• Monitoring is fairly simple, as long as it is done frequently and you understand the basics.

• Maintenance costs are very little once the system is set up.

• Plants grow much faster. This leads to larger plants with greater yields compared to soil based cultivation.

• yields of crops over the same cycle.

**Disadvantages Of Deep Water Culture Systems**

• Fluctuations in pH and nutrient concentration. This is a particular problem in small scale systems, where fast growing plants can lead to rapid changes in the pH and concentration of the nutrient solution, which can quickly cause problems for the health of your plants.

• Calibration difficulty. Again, in smaller systems, due to the low volume of nutrient solution, it is more difficult to accurately adjust the pH and concentration of the nutrient solution. Sudden swings in the characteristics of the nutrient solution due to imprecise calibration can again lead to negative impacts on your plants.

• The water temperature is difficult to keep within the target range as it will be quickly affected by the ambient temperature of the growing space.

• Constant oxygenation of the water of a deep water culture system is essential. The plant roots will not survive sitting in water that is not actively oxygenated. If your air pump has a failure or if there is a power supply interruption, this will very quickly lead to oxygen starvation and death of the plant roots.

These drawbacks, however, can be overcome by careful setup and maintenance of a deep water culture system. Once you have mastered the basics of setting up and monitoring a deep water culture hydroponics system, it is fairly easy to scale the system up and apply your knowledge to grow different plants.

## Common Questions About Deep Water Culture Hydroponics

### Nutrient Type

The nutrients required for a deep water culture system will not differ significantly from those used in other forms of hydroponics systems. The main determining factors will be the plants that you are growing and the stage of growth that the plants are at. Whilst people using hydroponics at a more advanced level or larger scale may wish to make their nutrient solutions up from the base nutrients, it is usually more practical for most people to use a range of mixable hydroponics products. Personally, I use the General Hydroponics Flora Series as it's tried and tested and allows for easy preparation and adjustment of nutrient solutions to suit almost all situations.

### Singular Versus Modular DWC Systems

Another common question is whether a single or a modular system should be implemented. Undoubtedly, it is safer, to begin with a singular system, in order to keep things simple until you increase your confidence. Once you have experience with the operation and maintenance of a deep water culture hydroponics system, you can go for a modular system to scale up your cultivation.

### Sterilizing The Water Solution

Inquiries about sterilizing the water solution are also quite common. Keeping the solution sterile has many advantages in that the solution will be more predicable and you don't need to worry about the problems associated with harmful bacteria and pathogens. However, a sterile system also loses the advantages of beneficial bacteria within the system.

A sterile deep water culture system takes discipline to maintain and involves additional sterilizing agents and anti-fungal preparations. Personally, for beginners, I wouldn't advise running a sterile system, but once you gain experience and look to create a larger scale deep water culture system, it is something to look into more closely.

### Temperature Monitoring

The temperature of the water reservoir is an important aspect of a deep water culture system. In general, the optimum temperature needs to be between 60°F (16°C) and 68°F (20°C), to make sure your plants remain healthy and able to absorb an optimal quantity of oxygen and nutrients. A higher temperature will reduce the amount of dissolved oxygen available in the water, causing plant roots to be at risk of drowning, even if it is being actively oxygenated. A decrease in temperature may trigger the plant into seasonal changes, negatively impacting the desired growth.

### Oxygen Level Monitoring

The role of regulating water temperature in preserving the oxygen content of the reservoir is particularly important due to the difficulty in measuring the oxygen levels continuously. Commercial oxygen meters are quite expensive to purchase, and low-end ones may not be very reliable. Monitoring the temperature and air flow are better techniques to maintain good oxygen levels rather than measuring precise oxygenation of the nutrient solution.

## Changing The Nutrient Solution

Another question that is quite common is how often growers need to change the nutrient solution. The answer to this question is subjective, as it depends on the plant type, stage of growth and the size of the water reservoir. In any case, a good guide is that a nutrient solution should be changed a minimum of every two to three weeks. Some planters may choose to re-adjust the nutrient balance of the existing solution instead of replacing it. This, however, is more difficult to control and may not give the desired results.

## pH And PPM/EC Adjustments

Regular monitoring and adjustment of the pH and PPM/EC of the nutrient solution is necessary to ensure that plants are able to grow to their full potential. Fluctuations outside the desired ranges can lead to nutrient lockout, deficiency or toxicity, which can lead to plant stress or death. The pH of the nutrient solution should typically be maintained between 5.5 and 6.5. During vegetative growth, it is better for the pH to be on the higher end of this range, whereas it needs to be on the lower end of the range during flowering stages.

If you are using a two or three stage nutrient solution, there will usually be guidance included within the packaging or on the bottles. Generally, I would advise caution with the PPM of the solution. It is usually better to err on the side of making a less concentrated nutrient solution, to ensure your plants will tolerate it. Over time, you can increase the concentration to levels closer to those recommended.

## Depth of Roots

It is essential to ensure the roots are submerged into the water solution, yet also make sure that the stem and foliage are exposed to the air. To stay safe, it is recommended to keep around 1 to 1.5" of the roots exposed to the air to make sure the stem is sufficiently far from the solution. At any rate, water bubbles will eventually reach these exposed sections of the roots and will prevent them from drying out.

## Plant Propagation

This can be done either by conventional means or by using an aeroponic cloner. Aeroponic cloners are quite simple to use and enables easy transplanting of the young plants, as they will have bare roots and can go directly into your deep water culture system.

## Best Plants for Deep Water Culture Hydroponics

The best plants that can be grown in deep water culture systems are those that do not need to flower. Lettuce and other herbs are particularly suitable due to the greatly accelerated growth rates which can be achieved compared to soil based cultivation. Kale, chard, collard greens, tomatoes and peppers are all excellent candidates.

## Building A DIY Deep Water Culture Hydroponics System

In order to build a Deep Water Culture system, you will need:

- Water container or reservoir
- Air pump
- Air hose and air stones for bubble formation
- Grow nets or baskets to hold the plants
- Growing media to support the plant in the basket
- Hydroponics nutrients
- Equipment to monitor pH and EC of the nutrient solution

The first step is to connect the air pump, the tubing and the air stone. Place the air stone at the bottom of the reservoir, with the tubing going out to the pump, which should be situated close to the reservoir.

A good way of suspending the grow nets or baskets is to cut a sheet of Styrofoam to the size of the top of the reservoir. You can then cut holes in this so that the net pots sit securely in these holes.

Next prepare your plants by putting them in the grow nets or baskets and secure them in place with your chosen growing media.

Next make up your nutrient solution and add it to the reservoir.

Finally, put your plants in place so that they sit with their roots well submerged in the nutrient solution. Ideally, you should maintain around 1.5" of the roots exposed to the air to avoid the risk of the stems becoming submerged in the water over time.

The water solution should have sufficient bubbles to resemble boiling water; these bubbles are necessary to oxygenate the water, to deliver sufficient oxygen to the roots to allow them to remain healthy.

The system will need close monitoring for several days after being set up to ensure that the roots are receiving sufficient water, and the pH and EC of the nutrient solution will need to be monitored carefully and adjusted as necessary.

## Types Of Aeration For DWC Systems

In order to introduce dissolved oxygen into the nutrient solution, two aeration techniques are used; namely air bubbles and falling water.

Air bubbles can be produced using the joint operation of an air pump and air stones. The air pump delivers air containing oxygen into the water through the air stone. These bubbles can also be formed using an air hose, which will produce a larger number of smaller bubbles. This increases the surface area of the bubbles, which increases the oxygenation of the water.

Instead of using air pumps and stones to introduce bubbles, some water culture systems may utilize falling water to aerate the nutrient solution. This is because as water turbulently falls into the water reservoir, it applies downward pressure on the water and allows more oxygen to dissolve into the solution through the exposed surface area. This technique is, however, quite uncommon for home systems as falling water is more suited to larger scale systems.

### Optimum Water Level in A Deep Water Culture Hydroponics System

The optimum water level in deep water culture reservoirs depends on the placement of the plants on top of the water surface. The question is, should the plant holders (whether it is Styrofoam or a basket or a lid with holes) be touching the surface of the water or slightly hanging above it?

As mentioned earlier, leaving an exposed portion of the roots on top of the water surface is healthy for plant growth to reduce the risk of root rot. It also allows a margin of safety to prevent the stems from becoming submerged. Stems and foliage will not tolerate even well oxygenated water like the roots are able to.

The height of the plant above the water surface also depends on the absorption capability of the growing media. If a highly absorbent medium is used, there is less risk of the roots drying out, and it is more acceptable to let the plants sit slightly higher above the level of the water. Hence, the absorption rate of the growing media needs to be considered when deciding the level at which to suspend your plants.

The size of the plant also determines the water level in the reservoir. If the roots of the plant are very short, the plant holder needs to be touching the water surface to make sure the roots get sufficient contract with the water solution and receive the oxygen and nutrients that are needed. When the plant holders are touching the water surface, the plant roots are better exposed to the nutrients needed which speeds up root growth during these early stages.

For larger plants with longer roots, it is more acceptable to submerge only part of the root structure in the nutrient solution, as the plant will still be able to obtain an optimal quantity of water and nutrients.

**Recirculating Water Culture Hydroponics Systems**

A recirculating water culture system is a modular deep water culture system that allows multiple smaller water reservoirs to be connected to a central reservoir, such that water is circulated from and to the central reservoir. This setup allows a deep water culture system to be scaled up efficiently, while only needing to maintain the pH and concentration of one central reservoir, rather than for multiple separate systems.

Recirculating deep water culture systems takes work as follows:

The air pump is connected to the central reservoir, which is then connected using a pump to smaller reservoirs or buckets that are also interconnected using similar pumps. In a series manner, as one reservoir is filled up, it overflows into the next one, and ultimately excess water is fed back into the central reservoir and is recirculated.

This technique is effective as it reduces the number of air pumps required for the aeration process, and allows control of oxygen and nutrients in the central reservoir rather than in each bucket separately. In fact, even if there is still an air pump connected to each bucket/container, their operation can be alternated or scheduled to reduce their on-time and reduce their costs of operation. The other advantage of this structure is that more air bubbles are introduced during the water filling process, and also falling water can be used as a supporting aeration system during the overflow of water from one bucket to another.

Deep water culture is a great method of hydroponic cultivation which is increasing in popularity, both at home and commercially. These systems are easy to build and require little maintenance when constructed correctly. Getting started may seem somewhat daunting at first, but with appropriate knowledge and the right guidance, you can get started with such a system without spending too much money. If you are thinking about trying hydroponics for the first time, this is a great option.

### Common Deep Water Culture Questions

What type of nutrients should I use in my deep water culture system?

Companies offer a variety of hydroponic nutrients, so it can be hard to figure out which is best for you. In my opinion, it's best to start out with something dead simple like the General Hydroponics Flora Series. It's a three-part hydroponic nutrient that you mix in varying quantities based on your plant's stage of growth.

### Should I use a singular or modular system?

If you're just starting out, go with a single reservoir setup. You can build them yourself or buy one of the many on the market. A modular DWC system is better for growers who know exactly what they want to grow and how much they want to grow. Start small and scale up as you get more experience.

### Should my reservoir be sterile?

This is not a yes or no question. Some hydroponic gardeners want to keep their reservoir sterile. This means they won't have any of the biological contaminants that might plague a hydroponic garden, like algae. But at the same time, they won't be able to take advantage of beneficial bacteria. If you do decide to add beneficial biology to your reservoir, just be aware that it comes with the risk of having not-so-beneficial biological organisms tag along for the ride.

## What should my pH and PPM / EC be for DWC?

Just because you're growing in a deep water culture system doesn't mean you need to adapt your pH and PPM / EC. The standard range that most plants prefer (pH 5.5-6.5) is fine, however you will want to customize and monitor this based on what stage of growth your plants are in. When they're putting on vegetation, you want to keep your pH in the higher end of that range, and when they're flowering, the lower end.

As far as your PPM / EC, don't blindly follow the feeding schedule on the back of your hydroponic nutrients. They are typically higher than necessary. Try cutting that amount in half and seeing how your plants respond. You can always adjust upwards quickly, while adjusting downwards is more challenging as your plants may have already suffered from nutrient burn.

## What should the temperature of my reservoir be?

This is one of the downsides of deep water culture: it can be hard to control the temperature of your reservoir. Aim for no higher than 68°F (20°C). If you get much higher, the oxygen level in your water starts to drop (even if you're oxygenating with an air pump and air stone).

Also try to keep it above 60°F (16°C). If it goes any lower, your plants think that they're moving into a new season, typically fall or winter. This means they'll start to divert more energy towards flowering, which you may not want.

## When should I change my nutrient solution?

The longest you should wait before changing out your solution is three weeks, but this is just a general case. It depends on:

- The type of plants you're growing
- The stage of growth they're in
- The size of your reservoir

If you want to avoid a complete change, you can try to add water with some nutrient solution mixed in to get the right balance again, but this is difficult to do. A complete change may be the better route.

## How do I know how much oxygen is in my nutrient solution?

Dissolved oxygen meters are available for sale, but they're pricy and might be overkill unless you want extreme precision. I wouldn't recommend purchasing a lower-end one though – they're not very reliable. The best way to "monitor" your dissolved oxygen levels is simply to do the things that ensure that levels are good, namely keeping the solution at the right temperature and running your air pump.

**How much of the roots should be submerged in my DWC reservoir and nutrient solution?**

First of all, make sure that only the root matter is submerged in your nutrient solution – no stem, and certainly no vegetation. You don't want to completely submerge the roots, either. I personally keep about 1-1.5″ of root above the water line. The bubbles from the air stone will typically pop and water will still land on the roots that aren't submerged, so you don't have to worry about them drying out.

**How would I propagate plants if I don't want to use a growing media in my DWC system?**

That's easy – use an aeroponic cloner. You'll save money on growing media and the plants that you propagate will have nothing but bare root when you transplant them into your DWC

**Are there any deep water culture specific issues to watch out for?**

Monitor your garden for the following issues, all of which are common in DWC systems:

• Root-related plant diseases like Pythium

- Rapid fluctuations in pH or PPM / EC / TDS
- Nutrient solution that is too warm

**How much faster do plants grow in a DWC system?**

Provided you're doing everything right, plants grown in a DWC system (or most hydroponic systems) will grow at least 15% faster. I have seen my lettuce grow almost twice as fast in my deep water culture setup vs. my outdoor garden.

**What plants grow best in a deep water culture system?**

The obvious answer is anything that doesn't have to flower. Many varieties of lettuce and lots of different herbs will work very well in DWC. They grow super-fast and healthy, making them a fantastic option. However, you can also grow tomatoes, peppers, and even larger fruits like squash...they just take a bit more effort.

**Are there any other tricks available to the DWC grower?**

A: Yes! DWC growers can easily manipulate the amount of moisture in the root zone. This, in turn, can trigger plant responses such as essential oil production, fruiting and flowering. A dryer root zone can increase essential oil production in aromatic crops such as basil and mint. (They do this as a means to conserve water.) A wetter root zone can cause plants to focus on vegetative production, particularly large fan leaves, which in turn speeds transpiration and photosynthetic potential.

# NUTRIENT FILM TECHNIQUE

While all hydroponic systems offer intriguing new ways to grow plants that our farming ancestors couldn't have even dreamed about, NFT hydroponics is the only one that depends on a constantly flowing stream of liquid to feed plants, with their roots dangling down as if dipping their toes in a gentle mountain stream.

Don't let the technical-sounding name fool you into thinking it's a difficult system to master. NFT hydroponics is simple to create and use, once you learn a few basic principles. So whether you're just starting out with hydroponics, or you have some experience but want to try out something different, read on, and find out how you can produce fast-growing plants with a hassle-free nutrient film technique hydroponic system.

## What Is NFT Hydroponics?

NFT hydroponics is a method of soil-less cultivation where plant roots are suspended above a stream of continuously flowing nutrient solution that provides them with all the water, nutrients, and oxygen they need to sustain rapid, healthy plant growth.

The term "nutrient film" refers to the ideal situation of having a constant shallow stream of nutrient solution passing over the roots. This ensures that only the bottom part of the root mass will be submerged in the nutrient solution, while the upper part is left exposed to the humid environment created inside the growing chamber, thus providing the roots with an abundant supply of oxygen.

Providing plenty of oxygen to plant roots is one of the key concepts behind hydroponics. While plants need carbon dioxide for photosynthesis, their roots need oxygen to facilitate the absorption of nutrients. And when their roots have greater access to oxygen, plants grow faster.

**How Does An NFT Hydroponics System Work?**

In nutrient film technique hydroponics, the chamber that the roots grow in takes the form of a tube-like channel that's set at a slight incline so the nutrient solution will flow through it. These systems often consist of multiple growing channels because there's a limit to how long a single channel can be for practical reasons and before nutrients start become depleted at the far end.

The plants are situated in holes in the top of the growing channel, so the roots are suspended above the nutrient solution inside the chamber while the crown extends above. There's often no need for a growing medium, with NFT hydroponics, other than net pots to support the plants.

The solution is pumped from a reservoir to the higher end of the growing channel, and after flowing through the length of the channel, it's returned to the reservoir.

Nutrient film technique hydroponics is therefore a closed system that recycles the nutrient solution, allowing you to conserve water and nutrients.

## Why Choose Nutrient Film Technique Hydroponics?

There are several reasons for choosing nutrient film technique hydroponics over other hydroponic systems, mostly having to do with its simplicity and ease of use. Here are some of its biggest advantages:

- Easy to build and maintain
- Easily adaptable to different spaces and plant requirements
- Can be built relatively inexpensively
- No need for growing medium
- Reduced need for aeration of nutrient solution in the reservoir, due to constant circulation
- No fussing with timers or watering cycles
- Uses less water and nutrients due to nutrient solution recycling

## What Are The Disadvantages Of Nutrient Film Technique Hydroponics?

One disadvantage of NFT hydroponics is its dependence on electricity, with the pump running 24/7. And if there's a power outage or pump failure, the roots will quickly dry out and your plants could be severely damaged, even if it's only a relatively short problem with the system.

The other disadvantage is that larger plants, with more substantial root systems, and fruit-bearing plants pose significant challenges for nutrient film technique hydroponics. These systems are best suited for smaller, fast-growing plants such as lettuce, leafy greens, herbs, and a few other plants that will be harvested before the roots grow large enough to fill the growing channel and block the flow of nutrient solution.

Fruiting plants are problematic for two reasons. First, they are heavy feeders, which will affect the pH and nutrient levels of the recirculating solution. Therefore, any time you have plants that are in their fruiting stage, you'll need to be very vigilant about monitoring the solution to prevent deficiencies. Second, it's been found that most fruiting and flowering plants do better when they are allowed to dry out between irrigation cycles, so the constant exposure to moisture that's provided by NFT hydroponics isn't the best environment for them.

### Nutrient Film Technique Hydroponics DIY Guide

To build a nutrient film technique hydroponic system, you will need these basic components:

- Growing channels
- Small baskets for holding plants

- Nutrient solution reservoir
- Submersible pump
- Tubing

Now let's walk through the details for setting up each component of an NFT system:

## Growing Channels

As we've seen, the growing chambers for nutrient film technique hydroponics need to be in the form of long, thin channels that are positioned on a slight incline, in order to create the shallow stream of nutrient solution that gives the system its name. (You'll find a discussion of flow rate and channel slope at the end of this DIY guide.)

PVC pipes may serve as the growing chambers, although these are not ideal, since their curved contour doesn't provide the optimal air-to-nutrient ratio for stimulating growth. The best materials for NFT system growing chambers are channels that are flat at the bottom. This way, you can create a nutrient film with a larger surface area for the roots to feed from while providing the upper part of the root mass with good exposure to the oxygen within the chamber.

The growing chambers for this system have holes all along the top for the plants. And ideally, the tops will be removable so you can easily monitor the nutrient solution, check for pooling and blockages in the flow of the liquid, and inspect the health of the roots. However, removable covers aren't a necessity for successful NFT system growing.

As for the ideal length of the growing channels, they should measure no more than 35-40 feet (about 10-12 meters). Any longer, and you risk depriving plants located at the far ends of the channels from where the solution enters, since roots have been absorbing nutrients from the flow of solution all along the way. Shorter lengths will also make your system more versatile in terms of space usage, allowing for system designs such as a single loop made of a series of channels, a circuit with channels in parallel alignment, and vertical systems.

There's one other reason to go with shorter channels: they're less likely to sag. When setting up your NFT hydroponic system, you need to take great care that your growing channels are correctly aligned to the desired slope and are well supported. It's a good idea to check the integrity of your system before you begin growing plants by running plain water through it. And you should continue to watch out for sagging during use, as this will disrupt the flow of the nutrient solution.

### Baskets For Holding Plants In An NFT System

Usually referred to as net pots or net cups, these small baskets are used in hydroponic systems that don't require any growing media as well as for starting plants. They hold the plants at the base of the stem and allow the exposed roots to grow and expand uninhibited within the growing chamber. With the exception of starting seeds or cuttings, the baskets in nutrient film technique hydroponics usually do not contain a growing medium, due to the danger of stem rot.

If you are careful about matching the size of the baskets and the holes they fit into, it will be possible to transfer larger plants to larger growing channels or even to different hydroponic growing systems as they mature and their needs change.

### Nutrient Solution Reservoir

As with any hydroponic system, your nutrient solution reservoir should be made of an opaque material, to prevent the growth of algae and bacteria.

If you locate your NFT system reservoir at the low point of the recirculating nutrient solution system, you won't need an additional air pump or air stone to enrich the nutrient solution with oxygen. As the liquid cascades from the return line situated a little ways above the reservoir, this action will constantly aerate the solution.

Also keep in mind that you will need to monitor the solution level in the reservoir as well as the pH and nutrient levels of the solution and make adjustments as needed.

## Submersible Pump

With nutrient film technique hydroponics, you need a reliable pump, as it will be continuously running to keep the solution flowing throughout the system. It doesn't need to be a high-powered pump, since the aim is to create just a thin layer of liquid flowing through the growing channels. So you'll want to choose a submersible pump with the lowest gallons-per-hour rating that has the capacity to pump solution from the reservoir up to the entrance of the first growing channel in the circuit, which will be the highest point in the system. From there, the solution will flow down through the circuit and eventually return to the reservoir, under the force of gravity.

Another consideration in choosing a pump for your NFT system is that you may want one that has an adjustable flow rate that you can easily modify according to the changing needs of your plants as they grow.

## Tubing

The amount of tubing you will need depends on the design of your system. A simple single-chamber NFT system will require, at the very least, enough tubing to deliver the solution from the reservoir to the high end of the growing channel, which could be set up to empty directly back into the reservoir.

## NFT System Flow Rate And Channel Slope Considerations

In an NFT hydroponic system, the growing channels are set at a slight tilt to create a rate of flow that results in a very shallow stream of nutrient solution. As a general rule, the flow rate in your channels should be around 1 liter per minute, which is about .26 gallons per minute or just shy of 16 gallons per hour. However, you can have a flow rate as low as .5 liters per minute or as high as 2 liters per minute without developing nutrient imbalances.

To achieve this range of flow rates, the channel slope ratio should be between 1:30 and 1:40, meaning that for every 1 inch (or every 1 centimeter) difference in height, you have 30-40 inches/centimeters in length.

Nutrient film technique hydroponic systems are easily adjustable to the changing needs of plants simply by modifying the flow rate of the solution through the growing channels. There are several ways to do this:

- Design your system so you can adjust the slope of the growing channels
- Design your system to include adjustable drains in the bottoms of the growing channels
- Use a pump with an adjustable flow rate
- Place inline valves at the entrances to the growing channels

## Starting Plants In an NFT Hydroponic System

This ability to modify the flow rate of your growing channels is particularly useful if you plan to start plants in your NFT system rather than starting them in a nursery and transplanting them after they have rooted.

To start your seeds or cuttings, you will need to use a growing medium in your baskets to support the plants and ensure they receive the moisture and nutrients they need. Choose a medium that's less likely to become waterlogged, such as oasis cubes, coco coir chips, or perlite.

To encourage rooting, raise the volume of solution flowing through the growing channels while reducing the flow rate. That is, you want the level of solution in the channels to be a bit higher, but the speed of the flow to be lower. Just make sure you don't raise the level too high, as you still need to have plenty of air in the growing chamber. If you raise the level, you should also increase the aeration of the liquid in the reservoir with an air pump and air stone to ensure you have the optimal amount of oxygen available to the newly forming roots.

This method of rooting can actually speed up the production cycle of your crop.

## Are There Nutrient Film Technique Hydroponics Systems Available To Buy?

There are a number of companies that manufacture and sell complete NFT hydroponics systems. Here are a few of them:

CropKing offers a variety of affordable NFT systems, including small introductory systems that are suitable for home hydroponics beginners and hobbyists, along with system components and growing supplies.

HydroCycle Hobby NFT Lettuce Systems are another good place for anyone new to nutrient film technique hydroponics to begin.

Although geared toward commercial growers, American Hydroponics offers several smaller versions of their sophisticated NFT systems for home growers and classroom settings as well as consulting services and custom designs.

# KRATKY METHOD

For aspiring indoor farmers interested in launching their own growing operation, it can be tough to decide how to get started. There are so many variables to consider and decisions to make. In addition, the time and effort involved, to say nothing of the initial financial costs, can be substantial.

Fortunately, there is an easy way for prospective farmers to explore indoor growing on a small scale without having to dive headfirst into launching a full-on commercial operation. Known as the Kratky method after the horticulturalist who developed it, this modest hydroponic system allows interested growers to learn and explore the basics of indoor growing, test potential markets, and work on a flexible schedule, all without having to commit to an expensive facility or a system requiring extensive maintenance.

## What is the Kratky method?

Developed by Bernard A. Kratky, a horticulturalist at the University of Hawaii, the Kratky method is a passive hydroponic growing method. In this context, "passive" means that no pumps, electricity, or moving equipment are needed during the growth cycle (this is in contrast to active hydroponic systems, which rely on electric pumps and other tools to recirculate the nutrient solution that feeds the crops).

### How does the Kratky method work?

As the simplest method of hydroponic growing, the Kratky method involves growing plants without soil simply by suspending them in a nutrient solution. In the Kratky system, a young seeding is placed onto a raft that sits on top of a tank or other container filled with water and a nutrient solution. The seedling's roots are suspended in the solution: from this, they draw up the nutrients needed to feed the plant. As the plant grows, the water level in the tank decreases, which creates an "air zone" that provides oxygen for the plant's roots. When it's time to harvest the plant, the nutrient solution will be almost completely depleted.

### What are the benefits of the Kratky method?

For growers, the main benefit of the Kratky method is its simplicity: the Kratky method offers the closest thing to a "set and forget" approach that a grower could hope to find. Seedlings start off suspended in all the nutrient solution they will need for their entire growth cycle, so the plants do not require any watering or feeding once they have been set up. Furthermore, there is no equipment to maintain, nor any moving parts that could break down. This allows new growers to keep things as basic as possible and to be away from their growing facility for days at a time (something that is not possible with most other growing systems).

### Who is using the Kratky method?

The simplicity and low cost of the Kratky method make it a great choice for aspiring farmers interested in exploring what indoor growing is all about without having to make a major investment of time or money. It's also suitable for growers who need to have a flexible schedule (for example, growers who are still working day jobs or who travel frequently) and for farmers who need to be able to easily manage a growing operation singlehandedly.

### What are some considerations and challenges of the Kratky method?

While the Kratky method is a great way for new farmers to begin experimenting with indoor growing, there are some challenges and considerations associated with this growing system, particularly if the intention is to scale up to commercial levels. They include:

Increased size, decreased efficiency – In the Kratky method, growing tanks must be thoroughly cleaned out after every three to five growing cycles. For a small-scale grower, cleaning out 10 or 20 tanks is not so difficult, but the larger the operation gets, the more laborious a task this becomes.

Inconsistencies may be more obvious – Since the Kratky method is so simple, any small inconsistencies in the system may be magnified in the final product. For example, if the ground underneath the growing tank is not level, or if the raft or liner that suspends the seedlings is not neatly installed, the water level in the growing tank will be uneven: too shallow on one end and too high on the other. This can result in seedlings drying up and dying due to lack of water, and, on the other hand, seedlings rotting due to oversaturation.

Quality material is key — Again, due to the simplicity of the Kratky method, the material used must be of high quality in order to achieve a good final product. For example, the raft material needs to be just the right thickness: too thin and it may become brittle and warped, too thick and it may not provide the growing seedlings with sufficient access to air.

In a traditional deep water culture setup, you typically have your plant in a net pot with growing medium and you place it in a reservoir. Then you fill the reservoir with nutrient solution up to a certain point, making sure it doesn't touch the net pot.

The airstone that you add to your system will create bubbles that pop at the surface of the water, hitting your growing media and feeding your plant's young root system. As the roots grow, they'll eventually hit the surface of the water and growth will explode from there on out.

With the Kratky Method, you actually fill your reservoir with properly conditioned water further, making sure to cover the bottom third of the net pot with water. The reason? Without an airstone, your plants will need water at the start of their lives, and this technique ensures they'll never dry out as your growing media is constantly wet.

As the plant continues to grow, it will use water and the water level will decline – but your plant's roots will have descended into the nutrient solution by that time.

You might be wondering, "Aren't airstones used for more than just wetting the growing media in the seedling phase?" You're right – this is where the beauty of the Kratky Method starts to shine. Because you are not refilling your reservoir, your plants will keep using up water and exposing more and more of their root systems to the air, which will ensure your plants get enough oxygen to survive and thrive.

## Potential Problems With The Kratky Method

Like any system, there are some flaws and considerations you should be aware of if you want to make sure you have a successful grow.

Good for Leafy Greens Only – This is designed to be a simple, hands-off method. That means it can't really account for the increased nutrient and water requirements of plants that bear fruit. Use this for leafy greens like lettuce, spinach, etc – not fruiting plants like tomatoes, cucumbers, etc.

Pests – Because your nutrient solution will be still (because you're not using an airstone), it can draw the attention of pests, namely mosquitoes. To avoid this, make sure that the reservoir is protected from any type of bug or pest, while allowing some oxygen and air to flow in as well.

Water Quality – You are not going to be replacing or adjusting the water level in your reservoir, so it's important to start with very high quality water. If you're going with the Kratky Method, I would recommend reverse osmosis or filtered water – get your PPM as low as possible so you avoid a dangerous concentration of salts.

Watch Your pH – If you're new to the method, you may want to pick up a pH pen and test it every day. Once you get the hang of how to prepare your nutrient reservoir for the plant that you're growing, you can leave the system to do its job!

# DRIP SYSTEM

The beauty of hydroponics lies in its flexibility. There is no right way to create a hydroponic system.

Depending on the available space, plant species and other variables you can use any one of at least half a dozen systems. Of these, one of the most popular and commercially viable options is a drip system.

## What Is The Drip Irrigation System?

A drip system is an active hydroponic system. This means that it uses a pump to feed your plants with nutrients and water regularly.

It is also called trickle, or micro irrigation system. As the name suggests, the system uses small emitters to drip the nutrient solution directly onto your plants.

A drip system is not unique to hydroponics. Such a setup is also widely used in outdoor gardens to deliver water and nutrients to individual plants.

It works equally well with soil as well as growing media. In fact, the drip system was initially conceived for outdoor cultivation of plants in Israel.

The system was developed to improve water efficiency in the outdoor cultivation of crops. It was later successfully adapted to hydroponics.

Instead of spraying or running water to the plants, the emitters secrete the liquid in a slow dripping action. This ensures that the system uses very less water.

You have a high level of control over the amount of water and nutrients supplied to the plants.

The system uses a network of feeder lines to deliver the water to the plants.This kind of setup is best suited for large growing operations.

This is the reason for commercial operations preferring drip hydroponics over other systems.

## How Does The Drip System Work?

The system usually uses individual pots for plants. The water from the reservoir is connected to the plants by a network of tubing.

There are two ways to apply pressure to the water supply. It can be a regular water pump or a gravity-based system.

Each individual plant gets at least one dedicated drip emitter. Each emitter has mechanisms that allow you to control the flow of water.

This adds to the overall versatility of the system; you can set different flow levels for various plants.

The flow to the plants has to be regulated in a drip system. The growing media needs to be given time to breathe in between flows.

If left uncontrolled, a drip system will flood the plants and eventually drown them.

So all drip systems use some kind of timer system to regulate the flow of water and nutrients to the plants. In typical situations, the pump is operated several times a day to send water to the plants.

Such a system requires considerable planning and effort in the initial stage. But once the drip lines and carefully installed, the system can run with minimal assistance.

These systems can be designed to have a high degree of automation.

## Variations Of The Drip System

Depending on your starting variables, you can rig up drip hydroponic systems in diverse configurations. But there are two main variants in drip systems, based on how you treat the excess water. They are:

### Recirculating/Recovery Systems

When the water is added to the hydroponic medium, all of it is not absorbed by the roots of the plant. In recovery systems, the excess water left behind in the medium is allowed to flow back to the reservoir.

This kind of a system is very popular for smaller, home-based drip hydroponic setups. While it is more efficient in its use of water and nutrients, the system also has some drawbacks.

When the wastewater is allowed to flow back to the reservoir, it affects the pH level of the reservoir water. This means you have to perform periodic maintenance on recovery systems.

The reservoir water will have to checked to ensure that the optimum pH and nutrient levels are maintained. This is easier and more cost-effective in smaller drip systems.

### Non-Recovery/Non-Circulating System

As the name suggests, in this system, any excess water is allowed to run off as waste. In usual circumstances, it is not very desirable to waste the water and nutrients like this. But since drip systems are highly conservative, the scale of wastage is relatively less.

This kind of system is very popular in larger commercial drip hydroponic setups. Commercial growers have the ability to use sophisticated timers to have maximum control over the water flow. This can keep run-off to a minimum.

There is another factor that makes such a system highly desirable for commercial growers. It requires lesser maintenance to the reservoir water.

You don't have to worry about the recycled water altering the pH and nutrient levels in the reservoir water.

### How to Setup A Drip Hydroponic System

As already mentioned, the drip system is very flexible. It can scale well according to the size and complexity of a growing operation. For a basic drip system, you will need the following essential items:

• Drip Emitters: Depending on the number of plants you plan to grow, you will have to buy an emitter for each plant. They are readily available at garden centers and hydroponic suppliers.

• Thin Tubing: spaghetti tubing is readily available in the market and is perfect for a drip emitter.

• PVC Tubing: These will be the main lines that carry the water and nutrients from the reservoir pump to the emitters. Depending on the size and complexity of your setup, the length and number of tubes required will vary. For smaller home-based setups, two-inch tubes are enough.

- Water Pump: a regular submersible pump is more than adequate for the task. Capacities of between 120-300 gallons per minute should suffice for smaller setups.

- A Tray: In smaller recirculating setups, you can get best results by having all the pots drain into a common tray. This is a simpler option than having separate run-off tubes from each pot to the reservoir.

- A Large Bucket/Bin: This will act as the reservoir. Choose between 10-20 gallon volume depending on the size of your setup.

- Small pots for your plant

- A Garden timer for the pump

- Aquarium grade silicone sealant

- A Hydroponic growing medium like coco coir

- A power drill, and hacksaw or PVC cutter to cut the pipes

The basic setup is simple. Place the pump inside the reservoir, and connect it to the emitters using the PVC and spaghetti tubing. Place individual drip emitters into the growing medium in each pot.

Don't forget to have adequate drainage holes in the pots.

Incidentally, you can also get the system to work without drip emitters. You can just poke holes in the spaghetti tubing and apply that directly to the growing medium instead.

But the emitters do provide additional control over the flow of water.

Place the pots in the tray, and then place the setup in such a way that the run-off drains into the reservoir. Give it at least several inches of height advantage over the reservoir for gravity to have an effect.

If the reservoir is in a higher location, you will need a pump to get the water back in.

Attach the timer to the pump power source, and set it to run on a daily schedule, like for 5 minutes thrice a day or something similar. It can vary depending on the water requirements of the plant.

### The Best Plants For A Drip Irrigation System

Since it gives you better control over the water and nutrient inputs, a drip system is ideal for a wide range of plants and herbs. It works well with different growing media as well, so this also increases the scope of this system.

The following are some of the plants you can grow with a hydroponic drip system:

- Lettuce
- Leeks
- Onions
- Melons
- Peas

• Tomatoes

• Radishes

• Cucumbers

• Strawberries

• Zucchini

• Pumpkins

Drip systems are considered especially suited for larger plants. These plants require larger growing media, which can retain larger amounts of moisture for a more extended period.

So, despite the slow watering system, the larger plants get proper hydration and nutrition in a drip setup.

For best results, a slow draining medium is preferred. The most popular options in this category include Rockwool, peat moss, or coconut coir.

Other media like clay pellets, perlite, and gravel can also be used successfully.

**Pros & Cons Of Drip System**

In hydroponics, a drip system has the following advantages:

• Provides more control over water and nutrient supply

• Flexible system that can be scaled for growth

• Requires low maintenance compared to other methods

- Affordable and cheap installation
- Less chance of system failure

It also has the disadvantages as follows, especially from a non-commercial perspective:

- Might be too complex for a very small grow operation
- If using water recycle system, maintenance is high (for reservoir water)
- If using non-recovery system, there is chance of waste

# EBB AND FLOW HYDROPONICS

Ebb and flow hydroponics, also known as flood and drain hydroponics, is often associated with aquaponics, since the growing medium can be used as a bio-filter for fish waste nitrates. But for plant growers who are not ready to bring live fish into their gardening routines, this type of hydroponics is great intermediate-level system that can support a greater variety of plants than other hydroponics systems can.

## What Is Ebb and Flow Hydroponics?

Ebb and flow hydroponics is a system of soilless plant cultivation that delivers water, nutrients, and oxygen to the roots of plants in cycles that consist of flooding the growing medium with water-based nutrient solution and then draining the liquid and allowing the medium to dry out.

The main principle behind ebb and flow hydroponics is that the intermittent flooding occurs in cycles that allow the growing medium to dry out between them. The plants benefit from this short dry period because it forces their roots to grow in search of moisture; and the more a plant's root system grows, the better it can absorb nutrients – and the faster and healthier the plant will grow!

## How Does an Ebb and Flow Hydroponics System Work?

The most basic ebb and flow hydroponics system consists of a growing container holding the growing medium and the plants, and directly beneath it, a nutrient solution reservoir, which contains a pump that's attached to a timer.

The pump turns on according to the settings of the timer, sending nutrient solution up to the growing container through a hole in the bottom of it. The level of the solution is regulated by an overflow tube, which empties into the reservoir below.

So the liquid in the growing container will rise to the level of the overflow tube and remain at that level as long as the pump continues to circulate solution up to the tank while the overflow tube sends solution back to the reservoir in equal measure. Then, when the timer turns the pump off, the liquid in the growing container is allowed to drain slowly back into the reservoir via the original tubing that brought it up.

There are several variations on this simple theme that allow you to create a system with multiple growing containers, one of which involves a second pump.

## Why Choose Ebb and Flow Hydroponics?

Flood and drain systems are popular hydroponics systems for several reasons, not the least of which is that they are quite easy to build. And because you don't need any pricey specialize components for a basic ebb and flow hydroponics system, you can put one together using inexpensive, everyday materials.

If you don't have a lot of space available, a simple ebb and flow hydroponics system can be quite compact, with the reservoir located right below the growing container. However, these systems are scalable and adaptable to different growing needs.

Another advantage is resource efficiency, since you are recycling the nutrient solution through the system rather than just running it through once then discarding it. This means you can let the automated system run itself for several days or weeks at a time, as long as it is set to meet the needs of your plants at their current stage of their growth cycle.

But perhaps the biggest advantage of ebb and flow hydroponics is that you can achieve success with plants that don't do so well in other hydroponic systems, such as cucumbers, beans, tomatoes, and other medium-to-large sized plants. Plants that are in their flowering and fruit-bearing stages respond particularly well to the practice of allowing the roots to dry out between irrigation cycles.

## What Are the Disadvantages of Ebb and Flow Hydroponics?

Ebb and flow hydroponics isn't for everybody. Here are a few downsides:

- Dependent on electricity and a pump

• Expanded systems are more complicated to build

• Unstable pH and nutrient levels of the recycled solution

• Need to watch for algae and pathogens in the open growing container

• Getting the cycle timing right can be challenging

## Is Ebb and Flow Hydroponics Suitable for Beginners?

Despite the easy setup, ebb and flow hydroponics is generally considered an intermediate-level hydroponics system. The basic principle of flooding, draining, and allowing the medium to dry out between cycles may seem simple, but in practice, it can be tricky to get the timing right, as it will depend on the ability of the growing medium to hold moisture as well as the changing needs of your plants.

Ideally, you want to allow the plants to dry out between flood cycles to reduce the risk of overwatering. However, determining the perfect timing of your flood and drain cycles will depend on a whole range of factors, including the growing medium, the stage and rate of plant growth, plant variety and the environmental conditions. Therefore, you will need to experiment a bit to get a good feel for the cycle timing, keeping in mind that plants consume more water and nutrients as they grow larger.

In general, you'll need to monitor the pH and nutrient levels of the solution, top up the reservoir as necessary, and change out the solution every week or two. Periodically flushing out the medium with plain water is also a good practice with any recirculating system.

## Ebb and Flow Hydroponics DIY Guide

These are the basic elements of a simple ebb and flow hydroponics system:

- Growing container
- Growing medium
- Nutrient solution reservoir
- Submersible pump
- Timer for the pump
- Overflow tube
- Solution inlet/outlet tubing

Here, you'll find a brief discussion of each of these components, followed by descriptions of the different types of flood and drain hydroponics systems.

## Growing Container

For the growing container in a basic ebb and flow hydroponics system, you can either place your growing medium and plants directly into the container, or you can have your plants in separate gardening pots and allow the growing media to wick up the solution from the bottom of each pot. The downside of having your plants all growing together in the container is that the roots can become entwined, making it difficult to remove the plants for transplanting or cleansing your medium of pathogens that can develop.

Regardless of which option you choose, your growing container will need to have two holes in the bottom: one for the inlet/outlet tube, and the other for the overflow tube. The holes should be at the far ends of the container from each other to ensure circulation of the nutrient solution for the period when the pump remains running after the system is fully flooded. You should also make sure the inlet/outlet tube can function properly as a drain, meaning you may need to have your growing container set at a slight tilt to ensure that this is the lowest point.

### Growing Medium

The growing media that work best with ebb and flow hydroponics systems are those that drain well, can support your plants, and don't hold too much moisture. Remember, the roots need to dry out between flooding cycles.

Clay grow stones are a popular choice, although rock-wool and rinsed gravel or sand can also work well. Or, consider using a combination of coco coir chips, to aid nutrient delivery during the drying part of the cycle, with a layer of river rocks placed at the bottom for drainage.

Perlite can also be an appropriate medium, but only if your plants are potted individually, as this material is too buoyant for the ebb-and-flow action of the liquid in a larger growing container.

## Nutrient Solution Reservoir

Your reservoir should be made of an opaque material, to avoid algae and bacteria growth. In a basic ebb and flow system, the reservoir is located directly below the growing container, so a shallower container will work best. The reservoir size will depend on the size of your growing operation and the type of plants you wish to grow.

You may need an air pump and air stone in your reservoir to make more oxygen available to the plant roots, if you are using a medium that doesn't provide great aeration, such as sand, or if your reservoir is not directly beneath the growing container, which allows the flow from the overflow pipe to serve up a little aerating action.

## Submersible Pump

Inside the reservoir, you have a submersible pump. By having it located directly underneath the growing container, you are minimizing the vertical distance that the nutrient solution must be pumped. So you can use a simple pond or fountain pump, as long as it has a strong enough flow rate to fill the growing container in a short period of time.

## Timer for The Pump

You don't need a high-accuracy timer for these hydroponics systems. A regular irrigation timer will do just fine, since the flood and drain cycles don't need to be timed to the second.

## Overflow Tube

The overflow tube determines the maximum level of nutrient solution in the growing container. In the standard ebb and flow system arrangement, it can simply be a single straight tube set at the height you want the solution to rise above the bottom of the growing container and allowing the liquid to fall back into the reservoir below.

However, you might want to place a T connector at the top that will pass liquid from the side and allow air into the pipe from the top to make sure it drains smoothly.

## Solution Inlet/Outlet Tubing

This tube leads from the pump to the bottom of the growing container and serves to both deliver the nutrient solution to the container and to drain the container after the pump is turned off. The diameter of this tube should be smaller than that of the overflow pipe, to ensure that you aren't pumping in more water than what flows out while circulating the liquid for a short time after the maximum level has been reached.

## Ebb and Flow Hydroponics System Variations

As mentioned above, there are several variations on the basic ebb and flow hydroponics system described in the previous section. They involve multiple growing containers and a second pump.

## Multiple Growing Container System (Single Pump)

A multiple growing container single pump system is designed with plumbing that delivers nutrient solution to each grow tray through piping running below the containers that has T connectors sending the solution up to the bottom of each one.

The containers will fill up with solution all at the same time, rising to the level determined by an external overflow pipe that bends back down and allows the liquid to pour back into the reservoir that's located below it. Since this type of flood and drain system depends on gravity to return the solution to the reservoir, the containers must be situated so they are higher than the reservoir.

## Surge Tank Design (Dual Pump)

This variation of ebb and flow hydroponics is meant for growing larger plants, as it allows for more vertical space by having the growing containers located on the floor rather than needing to be raised up above the reservoir. Instead of depending on gravity to drain the solution out of the growing containers, this system requires a second pump that's located inside a surge tank, which is a smaller container that will regulate the solution level across all the growing containers.

In this design, the nutrient solution reservoir contains a pump with a timer, as in the other systems; but here, the solution is sent into the surge tank when the pump is turned on. The surge tank feeds all of the growing containers from lines that enter them at or near the bottom, so the level of the liquid will remain the same across all of them. The liquid will rise until it reaches the level set by a float valve in the surge tank, which turns on a secondary pump that sends the solution back into the reservoir.

Both pumps will run for a short period to circulate the liquid, until the timer turns the main pump off. The surge tank pump, however, will continue to run, and the liquid level in the growing containers and surge tank will fall until it reaches a second float valve at the bottom of the surge tank that turns the secondary pump off. The containers have now been drained.

**Are There Ebb and Flow Hydroponics Systems Available To Buy?**

There are, indeed, several complete ebb and flow hydroponics systems available on the market today. Here are some of the top-rated ones:

Active Aqua's Grow Flow surge tank ebb and flow system is very versatile, efficient, and reliable.

The Viagrow Complete Ebb and Flow Hydroponic System is another good option for those who want to jump right in and get started planting.

# VERTICAL HYDROPONICS

What do you do when you don't have enough ground surface for all your planting needs? This is a problem that drove humans to develop the concept of vertical farming.

Think multi-storey buildings or skyscrapers, and you have the same working principle behind the concept. Vertical farming is all about cultivating more by stacking multiple layers of planting surfaces.

It is easy to see why the concept becomes highly desirable for hydroponics. Since indoor hydroponics enthusiasts often suffer from lack of floor space, vertical hydroponics is often the only choice.

## What Is Vertical Hydroponics?

Vertical farming is the growing of crops in vertically stacked layers. Vertical hydroponics, as the name suggests, is the combination of hydroponics and vertical farming.

So in a vertical hydroponics grow system, you will have several stacked levels, with plants being grown on each level. It is closely associated with gardening and farming in urban areas like cities.

The practice is also known by several other names. Tower hydroponics, tower gardens, vertical grow systems are the most popular names.

Incidentally, the practice of vertical gardening is certainly not new. It has its roots in Ancient history.

The Babylonians had a similar idea when they built the Hanging Gardens in around 600-500 BC. This Ancient Wonder had flowers, shrubs and even trees growing in massive tiered gardens.

In modern times, hydroponics and vertical gardening seem made for each other. Using soil as growing medium increases the weight of a vertical growing system.

Hydroponics, on the other hand, can reduce the overall weight of the upper layers by at least 30%, if not more. This means that you can stack more layers.

The main challenge then is in delivering adequate water+nutrients and light to plants at all the levels.

The Babylonians had to use water flowing from the mountains, along with manual water screws to irrigate the higher levels. Water pumps make this task much easier in the 21st century.

Providing light can be a challenge, especially indoors. If your hydroponics system is outdoors, then the ancient idea of using staggered layers is perfectly viable.

But staggering levels come at the cost of additional. Indoor vertical systems can use properly positioned grow lights to create an ultra-compact, and high yield grow system.

## Advantages of Vertical Hydroponics

Compact & Space Saving Design

This is one of the main reasons why many experts are touting vertical gardening as the future of food production. You can grow more produce even in small indoor spaces, making this ideal for urban farming.

## Does not require soil

This is, of course, a general advantage enjoyed by all hydroponics systems. But soilless growth is especially suited for a vertical system.

Lack of soil minimizes the growth of weeds or pests. It also makes vertical hydroponics the lightest and most practical form of vertical gardening.

## Efficiency & Productivity

With these systems, in the surface space required for one plant, you can grow at least 3-4, if not more. High growth can be achieved by using the right nutrient mix and proper lighting.

## Minimal Wastage & Maintenance

Vertical hydroponic towers typically have a closed nutrient+water flow system. There is no runoff as the water keeps circulating. This removes wastage of precious resources and nutrients. The whole process can also be automated to reduce maintenance.

## Disadvantages of Vertical Hydroponics

Of course, no system is without its weaknesses and flaws. There are several challenges that any grower will face when trying vertical hydroponics for the first time:

### Water Flow Challenges

It is much easier to design a grow system with just a single level of plants. Delivering water and nutrients equally to all plants is not a major headache here.

When plants are stacked, getting water to the top layers might require higher powered pumps. And unless carefully designed, the lower levels may get drowned.

### Light Supply Issues

Outdoor vertical tower gardens can mitigate the problem of light supply by using a staggered design. Instead of having layers directly stacked above each other, you can space them strategically.

But indoors, ensuring that all plants get the equal amount of lights can be challenging. Installing separate grow lights for each layer might the solution.

### Resource Intensive

Yes, you can grow more plants with vertical hydroponics. But that also means that input costs also scale accordingly.

Using motors and grow lights can entail mounting energy bills. The water will also need constant monitoring in closed flow systems.

But the advantages of vertical hydroponics outweigh these challenges in an era when land and cultivable soil is getting more and more scarce.

### How does a Vertical Hydroponic system work?

There are many different hydroponic techniques like ebb and flow and nutrient film technique (NFT). Due to the unique dynamics of a vertical system, NFT is often the easiest to do.

Nutrient Film Technique involves having a constant thin stream of water flowing over the root system of the plants. This is a closed, constant flow system, which makes it perfect for a vertical tower design.

### Vertical Hydroponic Tower

In a typical hydroponic tower, the idea is to use a tube system with a pump to get water to the top layers. From there, you can use the assistance of gravity to channel the flow down to the reservoir.

You can either use a single tube to deliver water to the top level or use multiple channels to different layers for optimal delivery of water and nutrients.

DIY designs typically involve the use of PVC pipes or thicker drainage pipes for the central tower. On these, smaller holes are drilled at intervals to house the individual plants.

The plants are usually housed in net cups to allow the water to flow through the root systems.

In a tower design, the plants are grown at an angle, typically around 45 degrees.

### Zig-zag Vertical Hydroponic System

Not all vertical hydroponics systems need to use the vertical tower design. Some outdoor designs use multiple PVC pipes arranged on a trellis frame at diagonal angles.

The pipes are usually in a compact zig-zag pattern going up. The plants are housed in net cups, placed in regular 90-degree angles.

These systems also use NFT techniques to grow the plants. The water with nutrients is pumped to the top pipe, from where it flows down in a constant stream.

With an indoor system, artificial lighting is a major concern. Since the plants in a tower vertical system are placed at an angle, the best option might be to use vertical grow lights.

Panels hanging from the ceiling may not be ideal since all the plants are at different heights. The best way to circumvent is by using multiple vertically mounted lights to cover all the growing surfaces uniformly.

## How To Set Up a Simple Vertical Hydroponic Tower

You will need the following ingredients for this project:

- A large PVC pipe (3" or greater diameter)
- End cap for the pipe
- A large bucket, minimum 5 gallons, with a lid
- Submersible pump
- Net pots, around a dozen
- Vinyl hose for the pump, ½ inch inner diameter
- Loctite PVC Epoxy
- LED Grow lights

You will also need the following basic tools:

- A miter saw (or a hand saw should do the trick)
- A finely bladed jig saw
- Tape measure
- A drill and various sized drill bits
- Pencil and paper
- Tape
- Ruler

The 5-gallon tub or bucket will be the reservoir. The PVC pipe will be placed into the tub by making a proper sized hole in the lid.

You can use any length of pipe depending on the height of the ceiling in your indoor grow area. For a standard tower, five feet seems like a safe choice.

Mark the PVC pipe with slots where the net cups will be placed. You can use the entire 360-degree surface area of the pipe for this purpose.

The holes drilled into the sides of the cup should be large enough to house one net cup. Plan the size of the net cups according to the size of the main PVC pipe.

You can cut up smaller PVC pipes to make holders for the net cups. Stick these to the sides of the main pipe to create 45 degree angled planters.

Use the tubing and motor to deliver water to the top of the tower. In a rain tower system, the water then percolates down through the inside of the pipe, delivering nutrients to all the plants.

What are the ideal plants for this method?

Fast growing plants and herbs are the best options for tower hydroponics. This includes leafy greens like:

- Lettuce
- Kale
- Chard
- Mustard greens
- Collard Greens

• Spinach

Other options include;

• Flowers

• Cabbage

• Basil

• Cilantro

• Mint

• Dill

• Chives

• Broccoli

Fruits and some veggies can also be grown. These include:

• Tomatoes

• Cucumbers

• Eggplants

• Peppers

• Strawberries

Vertical hydroponics is a godsend for folks who don't have access to the soil to grow things. Vertical grow systems can significantly increase the productivity of your indoor grow areas.

These days there are numerous commercially available readymade vertical grow kits on the market. They are a sure sign of the increasing interest in vertical hydroponics, especially among urban growers.

Increased input costs and complexity can be a drawback, but the benefits do outweigh these limitations. There is no doubt that vertical gardening is a vital technique for a more sustainable future.

# HYDROPONICS IN COMMERCIAL FOOD PRODUCTION

## Commercial Hydroponics

With the first successful application of hydroponics techniques in the 1930s the stage was set for a paradigm shift in crop production from conventional geoponics or cultivation in soil to hydroponics or soil less cultivation. The first crops to be commercially harvested with hydroponics included tomatoes and peppers, but the techniques were soon successfully extended to other crops such as lettuce, cucumbers and others. It was not long before hydroponics techniques were successfully adapted even to cut flowers production; in fact any plant can today be grown hyrdroponically.

## Commercial Systems Overview

Commercial hydroponics systems can be classified into bare root systems comprising nutrient film technique (NFT), deep flow and aeroponics systems and substrate systems.

Bare root systems do not use media to anchor the plant roots; the roots are left bare while in substrate systems plant roots are anchored in media such as perlite, vermiculite, sawdust, peat etc. Hydroponics is basically all about growing plants in a controlled environment and this is best provided outdoors in greenhouses that can incorporate several means to monitor, regulate and control the environment inside them. For instance, the air entering the greenhouse can be filtered to exclude entry to pests and parasites that can harm plant growth. Such means help provide optimal conditions for plant growth both in and out of season. In fact, hydroponics allows cultivation throughout the year which makes for year round availability of hydroponically grown produce at all major supermarkets across North America. Valued at 2.4 billion dollars the hydroponic greenhouse vegetable industry has a growth rate of 10 percent per year and accounts for nearly 95 percent of the greenhouse vegetables produced in North America.

The extension of the growing season is not the only advantage contributing to the growing popularity of hydroponics production with both growers and consumers. There are several additional advantages as well including nutritious, healthy and clean produce, improved and consistent vegetable quality and elimination of the use of pesticides and herbicides. Pesticides and other chemicals used in conventional agriculture have an adverse environmental impact; the run off from these chemicals contaminate groundwater supplies. Commercial hydroponics systems eliminate these toxic chemicals and contribute substantially to keeping the groundwater free from contamination.

**Yields**

Commercial hydroponics systems have proved more productive than conventional systems of agriculture not only in the laboratory but even in actual practice. Most commercial hydroponics greenhouse facilities are built large to take advantage of economies of scale; typically, these cover areas more than 10 acres while smaller ones measure around two acres. In the research greenhouse, yields with hydroponics techniques have averaged around 20 to 25% higher than in conventional soil cultivation. In actual commercial practice, however, over a number of years, the yield of hydroponically grown tomatoes can be more than double that of soil based systems due to the reduced turnover time between crops, better nutrition and crop management. Additionally, commercial hydroponics growing techniques are also less demanding of chemicals for root zone sterilization and control of pests, weeds etc.

The dramatic increase in yields with hydroponics is best illustrated if we consider the actual production figures of soil grown and hydroponically grown produce. Field grown tomatoes average yields ranging between 40,000 to 60, 000 pounds per acre; on the other hand, top growing hydroponics facilities in the US and Canada report average yields of more than 650,000 pounds of tomatoes per acre. Additionally, given the fact that only 10 years ago top hydroponics producers were producing around 400,000 pounds per acre, the increase in yields with improvements in growing practices has been truly phenomenal. Similar production figures can be quoted for other agricultural produce like cucumbers with 10,000 pounds per acre for field production and 200,000 per acre for hydroponic greenhouse yields. Hydroponics lettuce and pepper yields too average around four times the corresponding yields of agricultural production.

## Global Trends

Given the number of advantages of hydroponics it is not surprising that hydroponics techniques are increasingly finding favor for commercial food production in many countries all over the world. According to recent estimates countries having substantial commercial hydroponics production include Israel - 30,000 acres, Holland 10,000 acres, England 4,200 acres and Australia and New Zealand around 8,000 acres between them. The fastest growing area for commercial vegetable greenhouses is Mexico. There are several reasons for this including free trade and favorable winter conditions that attract vegetable growers in large numbers. Mexico has summers that are considered hot in the summer, but with greenhouses located at the right altitudes vegetables can be grown in the hot summers as well as the cold winters. Though much of the produce comes from low tech plastic houses, many of these greenhouses use hydroponics technology, which indicates the growing popularity of hydroponics in commercial food production.

# HYDROPONICS NUTRIENT SOLUTION

Hydroponics is one of the most direct ways of growing your crops, because hydroponic nutrients go through shorter processes before they reach plant areas that need their intervention to trigger growth. Hydroponic fertilizers contain the important nutrients your crops need for good growth rates. Nitrogen, phosphorus and potassium are the most basic nutrient elements. All plants need them in good quantities.

Your hydroponic crops need special hydroponic fertilizers because the latter are the only source of nutrients; there are no nutrients to be derived from soil and you are growing in organic or inorganic media that is devoid of natural nutrient content. Also, there are different types of fertilizers for different stages of plant growth and different kinds of plants.

Hydroponic fertilizers endeavor to give you a pleasant growing experience. The nutrient mixes are not complex to assemble and most fertilizers already contain the right amounts of groups of micro-nutrients. All hydroponics fertilizers are water-soluble.

Hydroponic fertilizers must be important sources of nitrogen. Ammonium nitrate in measured quantities is good for hydroponic crops. Potassium nitrate and calcium nitrate are also useful sources of hydroponic nutrients.

Hydroponic fertilizers are full of refined nutrients and impurities are few and far between. They are chemically stable and are easily absorbed by crops. In soil-based fertilizers, plant roots must sometimes search for traces of nutrient elements in the soil.

Hydroponic nutrients give the plant the energy it needs to grow at every stage of its life. Plants need different things at every stage. When they are younger, hydroponic fertilizers must act on increasing root density and doing things that will ensure the survival and strength of the plant. As they come closer to harvest time, hydroponic nutrients must affect growth and improvement - for instance, ensuring that the plant flowers well and produces big fruits. Thus hydroponic fertilizers must cater to the varied energy demands of a plant.

Specialized in nature, hydroponic nutrients cater to specific growing goals. There are different hydroponic fertilizers one could use to increase the root mass, grow bigger blooms and provide specific nutrients to your crops depending on the stage of growth they are in. pH is important because your hydroponic nutrient solutions cannot be excessively acidic or alkaline. Typically acidic, hydroponics nutrients bring the solution into the right pH range when mixed with tap water. Monitor the pH of your nutrients solution diligently so that your hydroponic crops stay healthy and get all the nutrients they need.

# HYDROPONIC LIGHTS

Hydroponic gardening requires three things for healthy plant growth. Water, hydroponic nutrients and special hydroponic lights. Most home hydroponic gardeners grow their plants either inside a greenhouse or in a room in their home or garage. Because this type of gardening requires less room than soil gardening you can grow more plants in a smaller space.

## Setting up the Hydroponic Garden

When setting up this type of garden the home gardener may choose to buy a special kit that has all the containers and the water system for his garden. In order to make sure that your garden is going to grow as it should you need to buy special hydroponic nutrients that contain not only the nitrogen, potassium, and phosphorous your plants need to be healthy but all the trace elements as well. The type of hydroponic lights you will need will depend a great deal on what you plan to grow.

## Hydroponic Lights

If you are planning on planting a small scale herb garden then you will want to choose fluorescent hydroponic lights for this type of garden. Florescent lights are placed close to the plants and aid in giving them the light they need in order to grow healthy and strong.

Vegetable and Fruit growth needs to use HID hydroponic lights. There are two types of HID lights used in this type of gardening. Metal Halide and High Pressure Sodium. The Metal Halide lights are used to encourage overall plant growth while the High Pressure Sodium is used to encourage the plants to flower and bear fruit.

There are now HID lamps that are designed so that you can use both the Metal Halide and the High Pressure Sodium bulb in the same lamp making it necessary to only have one lamp instead of two. You will want to look for a kit for Hydroponic lights that comes with reflectors to help direct the light where you want it to go.

**How Much Light Do You Need**

The amount of light you need will vary depending on the type of plant you grow. The amount of artificial light you will need for your garden will depend on where you live and whether or not you use a greenhouse or a room with limited outside light.

If you are using a greenhouse for your garden you will more than likely need very little artificial light during the late spring and summer months. However, you will need artificial hydroponic lights to supplement the sunlight during the fall or winter months especially if you live in an area where there is very little sunlight and frequent storms that keeps the natural light from reaching your greenhouse.

Some indoor rooms rely almost solely on artificial light as they have limited sunlight reaching the plants. To ensure that your garden grows and produces the way you would like, you simply have to give it the right amount of water, the proper hydroponic nutrients and the correct amount of light. Supplementing natural sunlight with hydroponic lights will ensure that your plants get the proper amount of light they need.

# PESTS AND DISEASES

Growing plants in your own hydroponic garden is hard enough as it is...

After all, you have to pick the right nutrients.

You have to get your water right.

And you might even have to build your entire grow room from scratch.

After all that hard work, when your plants finally start growing you may think you've made it because it looks like it's all downhill from here.

That's when "disaster" can strike.

You Are Under Attack!

You are now in a war against an army of insurgents struggling for dominion over your grow room.

Your enemy?

**Pests & Diseases**.

Hey, we've all been there.

No matter how well you try to follow the best sanitation practices in your grow room, pests and diseases are just a part of indoor hydroponic gardening.

We're Here For You

Remember, we don't just produce the best hydroponics nutrients here at Advanced Nutrients.

We pride ourselves on being a part of this community. We grew up in it, so it brings us mountains of joy and satisfaction to help other growers with tips, strategies, techniques and tactics for growing because we truly care about your success.

After all, you could be using the best nutrients in the world (Advanced Nutrients) and still see your harvest fail if you can't get rid of these common plant pests and diseases.

So let's get started...

**Most Common Indoor Plant Pests**

In order to win this war, you have to know your enemy.

That means being an expert at indoor plant pests identification.

Spider Mites – these tiny little buggers (less than 1 mm long) are probably the most common (and most hated) of all indoor garden pests. They are actually little arachnids and because of their small size you may not notice them until they do serious damage to your plants. There are two reliable ways to spot an infestation: one, look for spider-like webbing. Two, take a tissue and wipe gently on the underside of leaves—if it comes back with streaks of Spider Mite blood—you know you have mites.

Thrips – are also tiny (around 5 mm long). Though tough to identify, their damage is easy to see. Look for small metallic black specks on the top of leaves. This will often be accompanied by the leaves turning brown and dry (possibly with yellowish spots) because the thrips have sucked them dry.

Aphids – also known as plant lice. They can be green, black or gray. No matter what color, they can weaken your plants because they suck the juice out of leaves and turn them yellow. You may find them anywhere on the plant, but they often gather around the stems.

Whiteflies – are about 1 mm long and look like small, white moths. This makes them easy to spot, but they're harder to kill because they fly away when disturbed to a new plant. They also suck your plants dry causing white spots and yellowing.

Fungus Gnats – surprisingly, the adult fungus gnats are not harmful, but their larvae feed on roots and feeder roots, which can slow plant growth, invite bacterial infection—and if take to extreme—even plant death.

Now that you've learned a little about these hydroponic pests, let's talk about how to get rid of them.

## Ways To Fight These Dangerous Indoor Plant Bugs

There are quite a few methods to hydroponic pest control you can employ.

Use Sticky Traps – when you hang these around the room, you can trap the pets and that makes it easy to identify them (and of course, it takes them out of the game). Blue stick cards are good for thrips. Yellow cards attract fungus gnats and whiteflies. Tip: make sure some cards are at the soil/medium level of your plants—where fungus gnats congregate.

Various Sprays – you want to avoid chemical poisons like Avid or Eagle. And we can't vouch for the effectiveness of homemade sprays from household items like garlic, etc. But some growers have reported good results from organic "pesticide" interventions like

Azamax. Another non-toxic way to protect your plants—before an infestation—is to use Rhino Skin—a foliar application potassium silicate product that basically puts a coat of armor on your plants (protecting against pests & diseases). As a bonus, it maximizes resin glands too.

Beneficial Predators – some growers report success using beneficial predatory creatures like nematodes. Put these live predators into your medium and they can hunt down and kill the pests.

It's important to try out different strategies to see which ones work for you and your plants.

One thing to remember is that you can combine these strategies. For example, you can setup sticky traps while you apply Rhino Skin to help strengthen them against attack for a double-power preventative strategy.

IMPORTANT NOTE: Certain populations of these pests have grown resistant to various pest control methods in various parts of the country. To make sure your pest-fighting strategy is as effective as possible, you should talk to your local hydroponics store about what is working best on the pests in your area.

## Most Common Indoor Plant Diseases

To help you with indoor plant disease identification and diagnosis, here are some of the most common problems that occur ...

Powdery Mildew – Does it look like someone sprinkled white powder over your leaves and stems? It could be powdery mildew and if left untreated it will give you stunted plant growth, leaf drop and yellowing of plant tissue. If it gets too far, you'll lose your plant.

Downy Mildew – Don't get these two confused. Downy mildew mostly appears on the underside of leaves and doesn't look like a powder the way "powdery" mildew does. They both can cause yellowing of leaves which makes them easy to mis-identify though.

Gray Mold – Also called ash mold & ghost spot – you'll see it start out as spots on leaves that lead to fuzzy gray abrasions and will continue deteriorating until your plants are brown and mushy.

Root Rot – When you have too much water and pathogens in your medium/soil, you can get root rot. Plants will wilt and turn yellow. Roots can get mushy too.

Iron Deficiency – Plants lacking iron will lack chlorophyll, so you'll see the leaves turn bright yellow while retaining green veins. Sometimes this is misdiagnosed as some other type of disease when your plants are simply lacking iron.

Now that you know a little bit about hydroponics plant diseases, let's talk about how to heal your crops.

**How To Win Against Indoor Garden Diseases**

They say an ounce of prevention is worth a pound of cure and that's exactly right.

The best way to fight back against pests & diseases is to try to prevent them in the first place.

Make sure you're following these tips ...

Wear Clean Clothes – pests and diseases can "ride" into your grow room on you. So make sure you're wearing clean clothes when you enter your grow room. That includes shoes that you might have walked through disease/pest laden outdoor soils with.

Clean Up Spills, Runoff, Etc. – because so many molds and mildews and other disease problems can be caused by excess water/humidity, you'll want to make sure you police your water use.

Keep Plants Clean – make sure to pick up and clean and dead plant matter you find around. Also prune your plants when necessary to remove diseased leaves/branches. The less dead plant matter, the less pests/diseases. Guaranteed.

Hopefully, by now, you've realized the pattern here?

In short, cleaner is better!

How Proper Hydroponics Base Nutrients Can Help Prevent Plant Pests & Diseases In Your Indoor Garden

It may seem like common sense, but the healthier your plants are the less disease and pests you'll have to deal with.

That's why we recommend using Advanced Nutrients plant-specific nutrients for your indoor garden.

For example, keeping your pH in the hydroponics the optimal range (called the 'sweet spot') of pH 5.5-6.3 by using pH Perfect Base Nutrients will help prevent your plants from becoming weak and will help them fight off foreign invaders.

If you are growing in coco coir, then you need coco-specific base nutrients. That's because coco coir binds with iron & magnesium, which means your plants can become starved of these important nutrients if your base nutes don't contain optimal amounts of it.

That's why Advanced Nutrients has developed a full line of coco-specific base nutrients like pH Perfect Sensi Coco Grow A&B and pH Perfect Sensi Coco Bloom A&B that are specifically designed to contain extra (and exact) amounts of varied forms of iron for growing in coco coir.

## These Hydroponics Supplements Can Help Your Plants Health Too

Also, you should consider giving your plants supplements to keep them healthy...

As discussed, Rhino Skin is a soluble potassium silicate formulation that strengthens plants physically. It fortifies plant cell walls and protects crops from heat, drought, disease, and other stresses.

(A nice bonus is that this silicon also increases production of trichomes and essential oils for more potent, higher-quality harvests with greater market value.)

How does Rhino Skin shield your crops? Silicon mortars up the spaces between cell walls—and even coats plant cell walls themselves—thickening them and hardening them against attacks and other external stresses.

It also defends plants, increasing crop viability, and reduces the incidence and severity of powdery mildew, gray mold, and other pathogens, large and small.

**How To Get Your Plants Healthy Again After a Pest or Disease Attack**

If you're not using Advanced Nutrients base nutrients—with their dialed in calcium and magnesium levels—Sensi Cal-Mag Xtra is a necessity. Use it at the first sign of a Ca and Mg deficiency such as stunted growth and yellowing leaves.

And when your plants really get in trouble, reach for Revive. This concentrated combination of proprietary ingredients can turn around wilted, unhealthy plants fast. It contains rejuvenating substances that race into your plant's roots and leaves, get rapidly distributed throughout its vascular system and flood the plant with healing nutrition.

**Remember, Pests & Disease Are a Part of Growing. But You Can Be Prepared.**

The bottom line is that dealing with pests and diseases is just something that every indoor grower will encounter.

You will eventually be attacked, even if you've done everything right.

That's because fighting these foreign invaders is just a part of growing.

But knowledge is power and if you use these tips to prepare, you and your plants will be fine.

# MONITORING HYDROPONIC PRODUCTION SYSTEMS

Whether a grower is using a nutrient film technique or deep water raft hydroponic system, it is important to closely monitor pH, soluble salts and temperature to optimize plant growth.

Growers who are doing hydroponic production in nutrient film technique (NFT) or deep water raft systems should be monitoring pH and soluble salts content (electrical conductivity) more often than growers using container substrates.

"With hydroponics, especially with NFT production systems, the root zone conditions can change very quickly," said Neil Mattson, associate horticulture professor at Cornell University. "The pH can change very rapidly because the water doesn't have a lot of buffering capacity.

"With deep water culture (DWC) where there is typically a larger volume of water used, things like water temperature, pH, fertilizer strength and the overall concentration of the nutrients, are relatively stable over time as compared to NFT systems. In DWC, these parameters don't change that much hour to hour. There may be slight changes from day to day and more changes from week to week. Deep water raft systems don't generally take quite the degree of management that NFT systems do in terms of constant or continuous monitoring."

Mattson said it would be good for growers using deep water raft systems to monitor soluble salts and pH every day.

"In terms of taking action with deep water raft systems such as adjusting the fertilizer strength that can be done on a weekly basis," he said. "Adjusting the pH can be done daily or every two to three days. But that is better than with NFT systems that need continuous monitoring. Sometimes for nutrient management in NFT systems there is a need to do pH management every day if not several times a day. Some people have automated inline pH and EC sensors with peristaltic pumps that turn on automatically to add acid to the water reservoir or add fertilizer solution. Typically with NFT systems there is a much smaller water reservoir in relationship to the plant surface area that is growing."

Mattson said monitoring whether growing in a deep water or NFT system is especially important in the early stages of growth.

"The young plants are the most valuable because they are initially at a high density," he said. "The young plants need to get off to a good start because growers will never be able to recover that growth," he said. "If growers start with poor plants, they are never going to achieve the optimum plants they are trying to harvest.

Growers should focus on their crops more closely when they are younger."

Mattson said lettuces, leafy greens and herbs are the most common crops grown in deep water systems.

"I have also seen growers grow microgreens with raft systems," he said. "The microgreens are seeded onto substrate mats on top of the rafts. The growers add some weight to the rafts so the microgreens sit lower and are in constant contact with the water. This method has worked well for microgreen growers using pond systems."

**Maintaining water quality**

Mattson said deep water raft systems typically don't require as elaborate a water treatment system as NFT systems.

"There could be a benefit for water disinfestation for the raft systems, but growers in practice aren't really using that for a couple of reasons," he said. "A grower can't easily sanitize a whole pond at one time. All the grower can do is pump out water and run it through a disinfestation system and then pump it back in. A grower is never completely getting rid of all of the disease organisms.

"Some of the water is being taken out, treating it and putting the water back in and then taking up more of the pond water. A grower never fully gets rid of the disease organisms. More commonly with DWC, growers will periodically pump water out and sanitize a whole pond before refilling with a nutrient solution and transplanting."

Mattson said growers using hydroponic systems often have algae problems because algae will also access the water and nutrients.

"Algae make their own food," he said. "They photosynthesize and use light to make their own energy. Algae will grow and become established naturally wherever there is light, moisture and a source of nutrients. If light can be excluded from a surface this can help to deter algae formation. When sunlight hits uncovered pond water there is a food source for algae. This can occur whether a grower is using conventional or organic fertilizers. This can also occur with NFT systems if the channels aren't covered. If the channels are exposed to light where water and the fertilizer solution trickle down, algae starts growing very quickly.

"If light can be excluded from a surface this can help to deter algae formation. If a grower is using a pond system, he doesn't want to leave the pond water exposed to light. The water is covered with dummy rafts until that space is used again."

Mattson said growers who keep reusing the same pond water have found they don't normally run into problems with root diseases if temperature and dissolved oxygen are at optimal levels.

“There are communities of beneficial microorganisms that become established in the pond water that naturally suppress root diseases,” he said. “Even with the establishment of the beneficial microbes, growers need to maintain the dissolved oxygen level to near saturation (about 8 parts per million O2 at room temperature) in the pond water to keep the plant roots and beneficial microbes actively growing.

“Growers can bubble in air or can inject pure oxygen into the water. It is also important to circulate the pond water so there is a uniform gradient related to temperature, pH, fertilizer and oxygen.”

Mattson said the Cornell University Controlled Environment Agriculture group found good plant performance in a 1,500-square-foot pond where water was recirculated and distributed through manifolds in the pond. Pumping capacity achieved a complete water recirculation exchange every 12 hours.

### Monitoring water temperature

Water temperature can also be an issue with lettuces and leafy greens grown in warmer climates.

“The best water temperature is around 68°F so even if the air temperature increases it helps to delay bolting of lettuce and helps to reduce disease organisms,” Mattson said. “Water heats up much more quickly in a NFT system than in a deep pond system. The NFT channels are not insulated. The NFT water is in contact with a large surface area so it starts heating up quickly if the air temperature in the greenhouse is warm.

"A pond is usually well insulated. Often the outer edge and the floor of the pond will be insulated. There are also the polystyrene rafts that float on top of the pond so the pond does not heat up very quickly."

Despite having a beneficial microbial community in the water, Mattson said every once in while root disease can develop in the pond. Pythium is the major root disease.

"Usually it's because of warm water temperatures that occur under summer conditions," he said. "This can be a major issue for the grower who has to drain the pond, scrub and remove any debris, use a disinfesting agent and then refill the pond. The whole time the pond is being cleaned it can't be used for growing a crop."

Mattson said with NFT systems it is imperative to have a backup electrical source and pump backup because if there is an electrical outage or a water pump breaks then the plants can dry out within hours.

"In a pond if the power goes out, there is a concern about controlling the greenhouse temperature, but the plants are sitting in water and have access to plenty of nutrients," he said. "The supply of dissolved oxygen could become depleted or run out, but that would take days if not weeks for that to happen. It is a much more robust system in that way."

## Fertilizer considerations

Mattson said growers who are considering using organic fertilizers with either NFT or deep water raft systems need to be aware of issues inherent with the source of the nutrients.

"I have tried organic fertilizers in a pond system and found that biofilm grows very quickly," he said. "Organic fertilizers are byproducts of plants and animals. The biofilm microbes use the carbon in the organic fertilizers as a food source and use up a lot of the oxygen in the pond water. The microbes are respiring so it is difficult to maintain a good dissolved oxygen level in the water.

"The biofilm also quickly coats the plant roots making it more difficult for the plant roots to access oxygen and nutrients. They are not disease organisms, but the root system becomes coated with biofilm and the plants can't grow. The biofilm is starving the plants for oxygen and nutrients. In a pond, the biofilm, which is floating in the water, will also coat all of the surfaces in the pond including the walls and the rafts."

Mattson said another benefit of a NFT system in reducing biofilm buildup is the continual flow of water.

“There could still be biofilm and some coating, but the water in a NFT system is saturated with dissolved oxygen that is continually moving though the root zone,” he said. “That helps to deliver oxygen to the roots. There still may be some biofilm formation in the channels, but not nearly as much as in a pond.

“Growers who are using a NFT system and organic fertilizer are more used to starting a new crop over and over again. It’s up to the growers whether they want to start fresh with each crop cycle. Draining the reservoir after each crop cycle, cleaning the channels and the reservoir and sanitizing fits better with NFT systems. Growers using pond systems are not going to want to drain and clean the pond every crop cycle. That is very wasteful in terms of water and fertilizer and is labor intensive.”

# COMMON PROBLEMS ENCOUNTERED BY BEGINNERS

It is not difficult to get your hydroponic system set up. Most of the Hydroponic Gardening guides, especially those for beginners, include a section on build-it-yourself hydroponic systems. They provide a parts list, a tools list, and simple step-by-step instruction on how to build your own hydroponic system.

However, once the hydroponic unit is in operation, quite often, the beginners will discover problems, some may not be easily solved after the system has already been built. Therefore, while you are planning for your hydroponic system, it is always good to know the problems usually encountered by beginners. They may have influences to your requirements or ultimately design of your hydroponic system.

The followings are some of the examples.

**Problem#1**

There is a concern on just how much nutrients to be poured over the aggregate. Because for those hydroponic systems using a "light proof" container concept, you will not be able to see through the containers or down through the aggregate. So it is very difficult to gauge the amount or level of nutrient solution. Without this visibility, the plants may likely be killed by either under or overfilling.

The viable solutions can be either put a visual indicator showing the nutrient solution level or water sensors for automatic system.

**Problem#2**

The second problem is how often to pour nutrient over the aggregate. If you just follow the interval for your normal house plants in soil, for an example, three to five times a week, you would probably kill your hydroponic plants. For hydroponics, because of the wider air gap in the aggregate as compared to soil, the nutrient solution will tend to evaporate from the aggregate much more quickly than water from soil. So in general, you would need to supply nutrient to your plants at least once a day.

The simpler the system, the more frequently you will have to be around to add nutrient solution. The interval can be anywhere from one to four times a day depending on several factors, such as light, temperature, humidity, type and size of your plants, and the size of your container. This means that you cannot even go away for a weekend or your hydroponic plants would begin to suffer.

The solutions to this problem are either to get someone to "feed" the plants for you whenever you are away for more than a day or to have your hydroponic system automated.

## Problem#3

A third problem involves proper aeration (or supply of air or oxygen) for the plants' roots. This area usually is not a concern for soil gardening in the backyard because worms perform this function. In some hydroponic system, particularly those using PVC pipes with holes drilled for plants, too often the roots clog up the waterways and aeration in the root zone may become a problem.

Different systems will have different ways of providing proper aeration, for examples, using pumps, raised platforms or specific aggregate suitable for hydroponics.

Simple Problems?

To some people, these problems seem to be a matter of common sense. However, if you are new to this soil-less gardening concept and without going through the actual exercise once, you are likely to discover a lot of trivial problems like those mentioned if you do not have a good planning.